Peter Kennedy

Flow Analysis of Injection Molds

HANSER

Hanser Publishers, Munich Vienna New York

Hanser/Gardner Publications, Inc., Cincinnati

The Author:
Peter Kennedy, Moldflow Pty. Ltd. 259-261 Colchester Road, Kilsyth, Victoria 3137, Australia

Distributed in the USA and in Canada by
Hanser/Gardner Publications, Inc.
6600 Clough Pike, Cincinnati, Ohio 45244-4090, USA
Fax: (513) 527-8950
Phone: (513) 527-8977 or 1800-950-8977

Distributed in all other countries by
Carl Hanser Verlag
Postfach 86 04 20, 81631 München, Germany
Fax: +49 (89) 98 48 09

Library of Congress Cataloging-in-Publication Data
Kennedy, Peter (Peter K.)
Flow analysis of injection molds / Peter Kennedy.
p. cm.
Includes bibliographical references and index.
ISBN 1-56990-181-3
1. Injection molding of plastics. I. Title.
TP1150.K45 1995
668.4'12'0685--dc20 95-2593

Die Deutsche Bibliothek – CIP-Einheitsaufnahme
Kennedy, Peter:
Flow analysis of injection molds / Peter Kennedy. – Munich ; Vienna ; New York : Hanser ; Cincinnati : Hanser-Gardner, 1995
ISBN 3-446-18050-8

Printed and bound in Germany by Schoder Druck GmbH & Co. KG, Gersthofen

Preface

In recent years, the use of flow analysis software as a tool for designing plastic parts to be manufactured by the injection molding process has become widespread. In part, this is due to advances in plastic materials and demands for higher quality and reduced manufacturing costs. General acceptance by industry of the advantages of using flow analysis is evidenced by the number of commercial software packages available today (see Manzione [ML]).

Over the years, many descriptions of software for simulating injection molding have appeared in academic journals and books. While these were readily available to specialist readers, an understanding of the principles on which flow analysis software is based is difficult for the non-specialist to obtain. This is due to the fact that flow analysis software uses concepts from several disciplines. In particular, an understanding of thermal and rheological properties of plastics as well as numerical methods is necessary before the details of their application to the injection molding process can be appreciated. The recent publication of an excellent book by Tucker [TC] goes a long way to address this difficulty. Though not directly concerned with any commercial software, this book provides a thorough background for the interested reader.

Many Moldflow customers have requested a detailed description of our flow analysis software. This book is our response to these requests. Essentially it describes the basic scientific principles on which Moldflow flow analysis software is based.

In preparing the book the background of the likely reader was considered carefully. We wanted the concepts in the book to be accessible to any interested engineer. To achieve this, we have attempted to make the book as self contained as possible. An overview of the necessary material properties and introductions to finite element and finite difference methods are included. Of course it was not possible to treat these subjects in the detail given by specialist works. In fact our coverage of background material is restricted to that required for an understanding of its implementation in our software. It follows then, that the specialist reader may be disappointed in the level of detail presented. Nevertheless, it is felt that this compromise allows a wider readership and so is justified.

Although tensor notation is used in some Chapters, no detailed knowledge is required. All tensor quantities are expressed in appropriate coordinate systems as soon as possible. In general, the presentation of finite element methods given here is a little long winded compared to other books. This is to avoid specialist notations that, while more concise, may cloud understanding.

The Chapters that are specific to the way in which Moldflow model the injection molding process are:

- Chapters 5 & 6 which describe the assumptions made in the mathematical models for the filling and packing phase;
- Chapters 9 & 10 which deal with the numerical solution of the governing equations.

All other Chapters are included to support this material. It is therefore possible for the reader who is short on time, but has sufficient background, to go straight to these Chapters.

The descriptions of filling and packing are treated separately. It is possible to use the packing phase formulation given here for the filling phase also. The reasons for describing each phase are both historical and practical. Early versions of flow analysis software did not analyze the packing phase. Moreover much literature published today assumes the melt is incompressible. The description of filling given here is therefore still relevant despite the fact that the packing formulation offers some advantages.

In terms of the mathematical requirements, it is assumed that the reader is familiar with elementary differential and integral calculus and some matrix algebra. Specialized results are introduced as required. Certainly, there is nothing beyond what is covered in a typical engineering degree.

Finally it is a pleasure to acknowledge the help of my colleagues at Moldflow in preparing this book. Special thanks go to Gal Sherbelis and Hans Hatton for their assistance in editing and also Elizabeth Van der Peet who assisted with the typesetting, layout and the design of the book.

Contents

List of Figures

Chapter 1
Introduction

In injection molding of thermoplastics, molten material is forced into a mold where it cools and hardens. The injection molding cycle consists of a number of stages. In this chapter we establish some terminology for describing these stages. Detailed descriptions of the process and operation of injection molding machines, are available in several books (see for example [RR], [BM]) and will not be discussed here.

Although the injection molding process is apparently simple, there is a complex interplay between the processing conditions, mold geometry and material properties. The combination of these factors determines the final quality of the part. It is this complexity that forced the development of software to model the process.

1.1 The Injection Molding Cycle

Figure (1.1) shows a typical injection molding machine. In operation, plastic granules are placed in the hopper. As the screw rotates, granules are forced against the wall of the barrel and melt due to both the frictional heat generated by the screw and conduction from the heating units along the barrel. As the material melts it is conveyed to the front of the screw. During this time, the screw moves back to allow the melt to form a reservoir of material at the end of the barrel. Injection takes place when the screw moves forward as a ram and forces the melt into the mold. The pressure required to inject the material may be very high. This pressure acts on the area of the mold in contact with the melt and tries to force the mold open. The clamp unit attempts to keep the mold closed during injection by maintaining an opposing force on the mold.

Figure 1.1: Injection Molding Machine

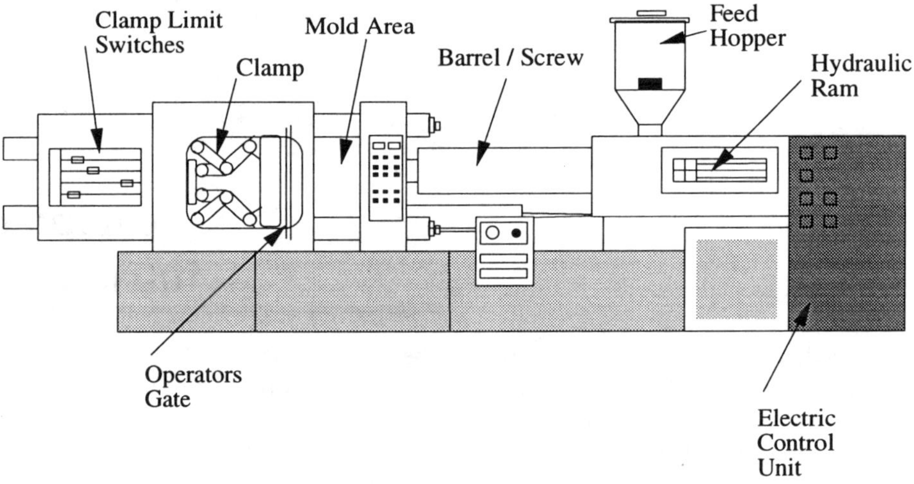

The injection molding cycle can be thought of as a series of distinct stages, each of which will now be described.

Let us assume that the cycle starts with the mold open and a full shot ready in the barrel.

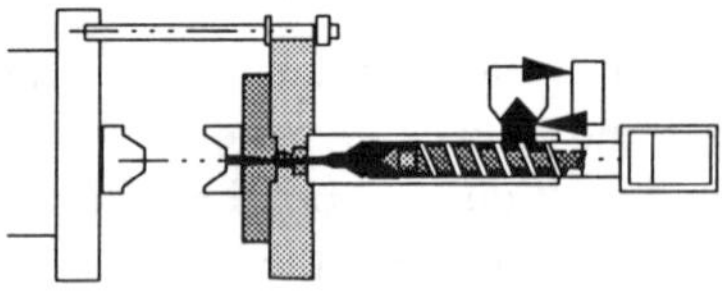

The first stage involves the closure of the die and the injection of molten material into the mold until the mold is just filled. This is called the *filling or injection phase*. The duration of this phase is called the *filling time*. During this phase the machine screw is not rotating, but simply acting as a ram to push the molten material into the mold. The ram moves forward at a rate that is defined by the molder.

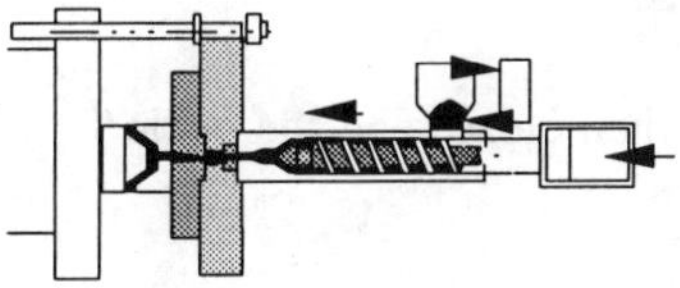

At the end of the injection phase, pressure is maintained at the feed system for a period of time called the *packing time*. This defines the second phase which is known as the *packing phase*. The packing phase is also called the holding phase. Application of pressure allows a little more material to enter the mold and so compensate for volumetric shrinkage of the material as it cools.

Eventually the packing pressure is reduced to a low value. This defines the start of the third phase known as the *cooling phase*. During this phase the material solidifies while in the mold.

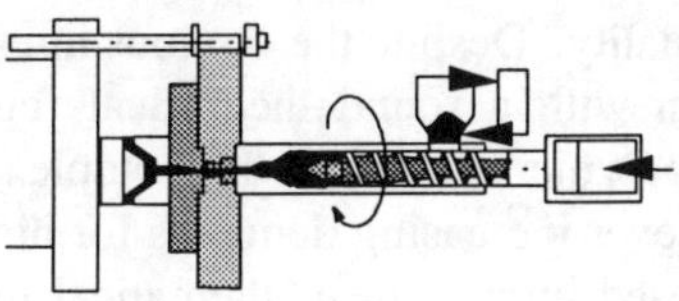

After the material is sufficiently solid, the mold opens and the part is ejected.

The mold then closes and the cycle starts again.

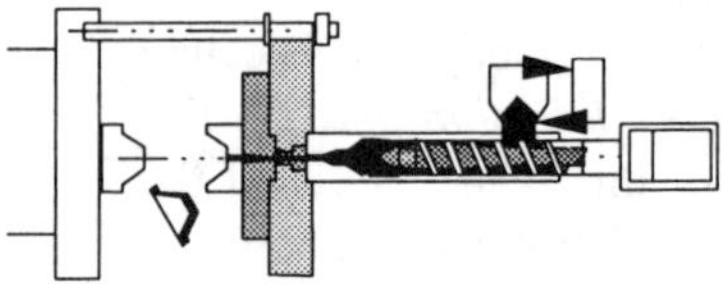

Moldflow flow analysis software analyses each of these phases. In this book we give a detailed description of the simulation of the filling and packing phases. The cooling phase is an integral part of the packing phase calculations and is not discussed separately.

1.2 The Need for Simulation Software

The conditions under which a part is molded have a significant effect on the quality of the component. Here we use the term "quality" in a broad sense to encompass such things as:

- surface finish
- dimensional stability
- mechanical properties

The relationship between processing and part quality has long been realized. However, due to the complexity of the interplay between the various processing factors and quality, injection molding remained something of an art for many years. An excellent overview of this may be found in Rubin's book [RI]. The bibliography of [RI] attests to the interest and dedication of many workers in the 1960's and 70's. There are hundreds of empirical studies cited; each adding to the general knowledge on relationships between processing and part quality. Despite the obvious interest, attempts to formulate an approach to plastic part design with a sound theoretical foundation were thwarted by the non-Newtonian nature of polymer melts and the complex heat transfer experienced in injection molding. However, ever increasing demands for higher quality injection molded parts in the 1970's saw increased interest in mathematical modeling of the molding process.

During this time, many pioneering studies were published (see for example [BI], [BG], [TB], [WL], [LW], [SJ], [KK]). These focussed on rather simple geometries and while of academic interest, offered little assistance to engineers involved with the day to day design of plastic parts and molds.

In 1974 Moldflow Pty. Ltd. offered commercial software on a world wide computer time sharing system. This software allowed users to determine processing conditions (melt temperature, mold temperature and injection time) and to balance flow in cavities and runner systems. Although accepted at the time, the software was quite difficult to use as the cavity geometry had to be flattened out to form what was then called a "lay flat" [AC].

A considerable break through was made with the introduction of finite element methods by Shen and Hieber [HS]. Though not yet practical, their use of finite element methods indicated the advantages of the technique for mold filling analysis. In 1983 Moldflow released a finite element flow analysis program for analysis of three dimensional geometries. This was the first commercially available package and found ready acceptance in the market. For the first time, designers could model the part under consideration in a form which resembled the component and display the simulation results on the model. Early versions of the software modeled only the filling phase. Analysis of the packing phase was developed as part of Moldflow's research into prediction of shrinkage and warpage. In this book we describe both the filling and packing phases.

Chapter 2
Material Properties

Analysis of the filling, packing and cooling phases requires data on the physical properties of the plastic material and the material from which the mold is constructed.

In this chapter we discuss some general properties of polymers and the data required for flow analysis software.

2.1 Types of Polymers

The molecular structure of polymers used in injection molding directly influences the physical properties of the material. For our purposes, a polymer molecule may be regarded as a long chain of covalently bonded units with additional chemical groups forming branches along the primary chain. This structure leads to two basic types of materials: thermosets and thermoplastics.

Thermoset materials become fluid on heating but as the temperature increases, a reaction known as crosslinking causes molecules to bond at the branch groups along the chain. This process is called *cross-linking* and leads to the formation of a rigid three dimensional lattice structure. Once the cross-linking occurs, the material will not remelt on heating.

The molecules in a *thermoplastic* material are not chemically joined. These materials can be made fluid by heating and solidify when cooled. Moreover this process can be repeated any number of times provided the material does not undergo any permanent chemical change due to the applied heat.

Because both thermosets and thermoplastics soften when heated, they can be used in injection molding. The main difference in processing is the mold temperature. For thermoplastics, the mold is run "cold", whilst for thermosets the mold is "hot".

Thermoplastics will be the principle focus for this book. Most of the material properties and numerical methods described later are however applicable to thermoset materials also.

2.1.1 Amorphous and Semi-crystalline Materials

The bonding between atoms along the main molecular chain is flexible in the sense that segments of the chain can rotate about the bonds. In the melt state, molecules have increased mobility and tend to adopt an entangled, disordered configuration. If the temperature is lowered, and in the absence of external forces the molecular configuration remains entangled, the material is said to be *amorphous.*

As an amorphous polymer in the melt state is cooled, it becomes rubbery. Further cooling will cause the material to change from a rubbery state to a hard, glass-like material. The temperature at which this transition occurs is called the *glass transition temperature*, T_g. In actual fact the transition is not as sharp as the above suggests. Typically the transition occurs over a temperature range of 10K.

While for amorphous polymers, molecular entanglement is preserved as the melt cools, it is possible that molecules may align themselves in a regular lattice shape. Such materials are said to be crystallizable. These materials frequently possess both amorphous and crystalline regions and so are called *semi-crystalline*. It should be noted that the lattice structure here is not due to chemical bonding as in the case of cross-linking.

2.2 Viscosity, Shear Rate and Shear Stress

Imagine a fluid, at fixed temperature, confined between two plates that are a small distance H, apart. Suppose the bottom plate is stationary and the top plate moves with a constant speed v_x in the x-direction due to application of a force F as illustrated in Figure (2.2). Such a flow is called steady simple shear flow.

Figure 2.2: Steady Simple Shear Flow

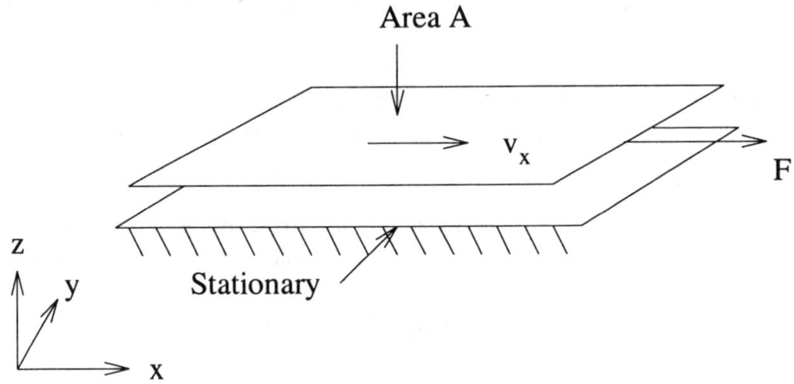

The *shear stress* τ, acting on the top plate is given by

$$\tau = \frac{F}{A} \quad \textbf{(2.1)}$$

where A is the area of the plate. The dimensions of shear stress are force per unit area. It has been determined empirically that the shear stress is proportional to the velocity gradient of the fluid, dv_x/dz. That is,

$$\tau_{zx} = \mu \frac{dv_x}{dz} \quad \textbf{(2.2)}$$

The velocity gradient dv_x/dz is called the *shear rate* and is denoted by $\dot{\gamma}$. It has units of reciprocal seconds. If we think of the flow as composed of a number of layers, each moving at a different speed, the shear rate is a measure of how fast the layers slide over each other. Therefore,

$$\tau_{zx} = \mu \dot{\gamma} \quad \textbf{(2.3)}$$

The constant of proportionality μ, is called the *viscosity* of the fluid. It relates the shear stress to the rate of deformation of the fluid. Later we will extend the definition of shear rate to handle more complex flow fields. For these more complex flows it is necessary to use the stress tensor and rate of deformation tensor in the definition of a Newtonian fluid. That is, Eq. (2.3) becomes

$$\underline{\tau} = \mu\underline{\dot{\gamma}} \qquad \textbf{(2.4)}$$

2.3 Newtonian and Non-Newtonian Fluids

Fluids for which stress is directly proportional to shear rate, that is Eq. (2.3) holds, are called *Newtonian.* Note that for such fluids, the viscosity is constant at a given temperature. The viscosity may vary with temperature however.

Fluids for which Eq. (2.3) does not hold are called *non-Newtonian.* Molten polymers are non-Newtonian. Moreover they exhibit *viscoelastic* effects. The current Moldflow flow analysis software does not account for these effects, and so they are not discussed here.

In order to model the flow of molten polymers, a modification to Eq. (2.3) is made to allow the viscosity to be a function of shear rate. Therefore Eq. (2.4) becomes:

$$\underline{\tau} = \eta(\dot{\gamma})\underline{\dot{\gamma}} \qquad \textbf{(2.5)}$$

Where $\eta(\dot{\gamma})$ is called the *viscosity function.* In what follows we will simply call this the viscosity.

In Chapter 3 we will discuss the generalized Newtonian Fluid in more detail.

2.4 Viscosity Models for Flow Analysis

To model the injection molding process, a viscosity function (or model) is required. A number of well known models are available. It is important to choose a model which is both accurate over the processing range and for which data can be readily be obtained. Some common viscosity functions are:

- Power law models (Moldflow first order model)
- Moldflow second order model
- Moldflow matrix data
- Ellis model

- Carreau model
- Cross model

Some plots of viscosity versus shear rate at several temperatures are shown in Fig. (2.3). These curves are typical for materials commonly used in injection molding. It should be noted that some materials, in particular liquid crystal polymers and some engineering blends, do display different characteristics. The aim of the viscosity model is to match the observed behaviour of the material as closely as possible.

Figure 2.3: Graphs of Viscosity versus Shear Rate at Several Temperatures

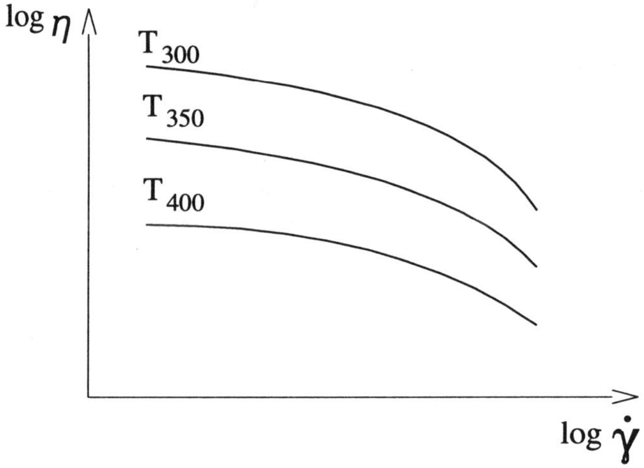

2.4.1 Power Law Model

This model has the form:

$$\eta = m\dot{\gamma}^{n-1} \tag{2.6}$$

where m and n are constants. When $m = \mu$ and $n = 1$ we obtain the relationship for a Newtonian fluid i.e. Eq. (2.4). For polymer melts, the constant n lies between 0 and 1.

Taking natural logarithms of both sides we obtain:

$$\ln(\eta) = (n-1)\ln(\dot{\gamma}) + \ln(m) \tag{2.7}$$

A plot of $\ln \eta$ against $\ln \dot{\gamma}$ shows this relationship to be linear. Therefore the power law model can represent the behaviour of polymer melts in the high shear rate region. It is also quite easy to fit experimental data to this model and so determine the constants m and n.

The main disadvantage of the model is in the low shear rate range. Here it is incapable of predicting the Newtonian viscosity, η_0, which is defined as the viscosity at zero shear rate. Despite this disadvantage, the model has been widely used for modeling flow in injection molding. In the filling phase particularly, shear rates are frequently high enough to justify use of the first order model.

Temperature dependence is usually included by multiplying Eq. (2.6) by an exponential term. That is,

$$\eta = m\dot{\gamma}^{n-1} \exp{(cT)} \tag{2.8}$$

where c is a constant and T is temperature. Equation (2.8) is the first order viscosity model available in Moldflow flow analysis software.

2.4.2 Moldflow Second Order Model

In order to improve the modeling of viscosity in the low shear rate region, the following model has been used by Moldflow.

$$\ln \eta = A_0 + A_1 \ln \dot{\gamma} + A_2 T + A_3 (\ln \dot{\gamma})^2 + A_4 T \ln \dot{\gamma} + A_5 T^2 \tag{2.9}$$

where the A_i are constants.

This model has been criticized on the grounds that it is based on empirical observations. However, this is of little relevance for our purposes. What is required is a model that fits the experimental data as closely as possible and the second order model is able to characterize the observed behaviour of polymer melts. In particular, the second order model can model the following features of polymer melts:

a. convergence of iso-shear rate curves with increasing temperature
b. convergence of isotherms with increasing shear rate
c. the increasing rate at which viscosity decreases as shear rate is increased
d. the decreasing rate at which viscosity decreases with increasing temperature.

There are, however, some disadvantages to this model. The most important of these concerns the fact that although very flexible, the model can introduce behaviour that is not rheologically sensible. For example, it is possible [GB] that $\partial\eta/\partial\dot{\gamma} > 0$. This means that viscosity increases with increasing shear rate and is clearly not realistic. Such behaviour requires that care be taken when fitting the experimental data to the model, to ensure that suitable cut offs are used at the low shear rate range, to prevent this occurring.

2.4.3 Moldflow Matrix Data Model

This model is simply a collection of triples (viscosity, temperature, shear rate) obtained by experiment. No functional dependence is assumed so no curve fitting is done as in the case of first and second order models. The analysis program linearly interpolates between the data points closest to the existing set of conditions.

The matrix data model is suitable for materials with unusual viscosity characteristics such as liquid crystal polymers. Neither the first or second order models, nor the models mentioned below, can adequately describe the viscosity behaviour of these materials.

For effective use, matrix data must be measured over as wide a range as possible. Other models that impose a functional relationship can be extrapolated outside the range of measured data, though results can be inaccurate (see Section 2.6). Matrix data can not be extrapolated to the same extent as these models, as only linear extrapolation is possible.

2.4.4 The Ellis Model

The Ellis model expresses the viscosity as a function of shear stress τ. It has the form:

$$\frac{\eta_0}{\eta} = 1 + \left(\frac{\tau}{\tau_{1/2}}\right)^{\alpha-1} \tag{2.10}$$

where $\tau_{1/2}$ is the value of shear stress for which $\eta = \eta_0/2$ and $\alpha - 1$ is the slope of the graph $\ln[(\eta_0/\eta) - 1]$ versus $\ln(\tau/\tau_{1/2})$.

2.4.5 The Carreau Model

The Carreau model has the form:

$$\frac{\eta - \eta_\infty}{\eta_0 - \eta_\infty} = \left[1 + (\lambda\dot{\gamma})^2\right]^{(n-1)/2} \tag{2.11}$$

where η_∞ is the *viscosity at infinite shear rate*, η_0 is the *viscosity at zero shear rate*, n is a constant with the same interpretation as that in Eq. (2.6) and λ is a *time constant.*

2.4.6 Cross Model

The Cross model has the form:

$$\eta = \frac{\eta_0}{1 + (\eta_0 \dot{\gamma}/\tau^*)^{1-n}} \tag{2.12}$$

where τ^* is the shear stress at the transition between Newtonian and power law behaviour.

2.5 Viscosity Model Requirements

The most important requirement of a viscosity model is that it represent the observed behaviour of polymer melts. In particular, a viscosity model must meet the following requirements:

- In viscosity/temperature space:

  a. the viscosity should decrease with increasing temperature
  b. the curvature of the iso-shear rate curves should be such that the viscosity decreases at a decreasing rate with increasing temperature
  c. the iso shear rate curves should never cross over.

- In viscosity/shear rate space:

  a. the viscosity should decrease with increasing shear rate
  b. the curvature of the isotherms should be such that the viscosity decreases at a decreasing rate with increasing shear rate
  c. the isotherms should never cross over.

2.6 Extrapolation of Viscosity Data

Regardless of the chosen viscosity model, problems may occur in flow analysis. This is because the range of temperature and shear rate over which viscosity is measured in experiments may not cover the range experienced in the simulation. In particular, as the simulation must include the effect of solidification during fill and packing, it is necessary to have viscosity data at temperatures at which the material is solid. Unfortunately, it is not possible to measure viscosity at such low temperatures. Some extrapolation from the measured data is therefore necessary.

The lack of measured data at low temperatures is particularly problematic for semi-crystalline materials. These materials exhibit a sudden change in viscosity as they pass from melt to solid. This rapid transition is not evident in viscosity functions that are fitted to experimental data. Extrapolation to low temperatures will therefore not give accurate results in simulations.

Figure 2.4: Extrapolation of Viscosity Data at Low Temperature

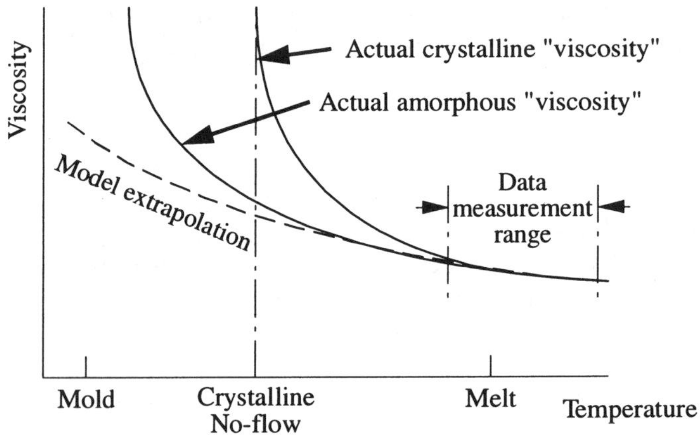

Figure (2.4) illustrates the problem. From the Figure it can be seen that extrapolation gives relatively low viscosity values even at the temperature of the mold. Extrapolation of typical viscosity models will therefore underpredict viscosity at low temperatures. It follows that flow analysis software using this extrapolated data will under-predict pressure.

Moldflow software uses the concept of *no-flow temperature* to overcome this problem. The no-flow temperature is discussed later in Section 2.7.3. For now we will define it to be the temperature below which the material can be considered solid. The concept of no-flow temperature, when coupled with a viscosity model based on data measured over a wide range of shear rate and temperature, gives excellent results in simulations. Figure (2.5) shows a viscosity model based on the Moldflow second order model with no-flow temperature. Note that at temperatures below the no-flow temperature, the viscosity is assumed infinite.

Figure 2.5: Viscosity Data with No-Flow Temperature

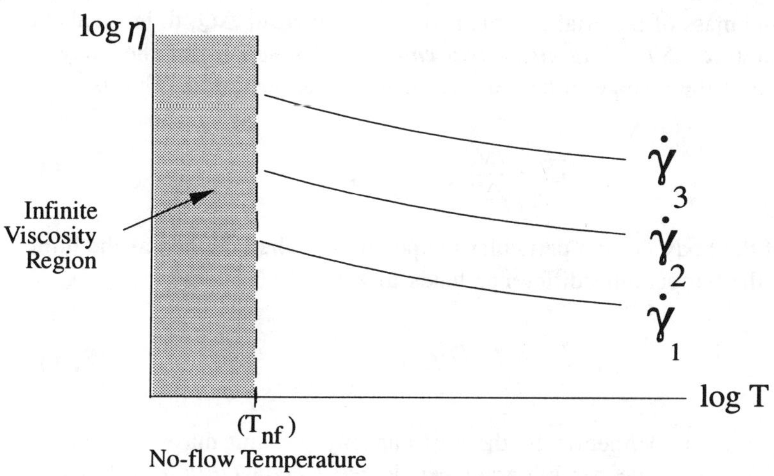

2.7 Thermal Properties for Flow Analysis

Heat transfer in the injection molding process requires attention to three fundamental phenomena:

- convection due to incoming melt;
- conduction out through the mold to the cooling channels;
- shear heating caused by the deformation of the melt as it flows into the cavity.

Modeling of these thermal effects requires material data such as:

- specific heat;
- thermal conductivity;
- no-flow temperature.

In this section we discuss these properties.

2.7.1 Specific Heat Capacity

Suppose we have some mass of material to which we add some heat, ΔQ, to bring about an increase in temperature, ΔT. The *mean heat capacity*, denoted $\overline{C}$, of the body is defined to be the ratio of the change in heat to the change in temperature. That is,

$$\overline{C} = \frac{\Delta Q}{\Delta T} \tag{2.13}$$

The *heat capacity* of the body C, at a particular temperature, is then defined as the limit approached by $\overline{C}$ as the temperature difference tends to zero. That is,

$$C = \lim_{\Delta T \to 0} \frac{\Delta Q}{\Delta T} \tag{2.14}$$

The *specific heat capacity* is defined to be the heat capacity per unit mass of material. The unit of specific heat is Joule per kilogram per degree Kelvin (J/(kg.K)). Specific heat capacity may be measured under conditions of constant volume or pressure and is denoted by c_v or c_p respectively. Due to the large stresses exerted on the containing vessel when heating a sample under constant volume, the use of c_p is most common.

Specific heat determines the amount of heat required to melt the material and therefore the amount of heat that must be removed after molding. It is used to calculate the heat lost due to conduction from the melt into the mold during filling and packing and for determining the temperature rise due to shear heating as the melt flows into the mold.

In general the specific heat of polymers is much higher than that of metals. Table (2.1) gives some values for common polymers and steels used in injection molding. The polymer values are from [RF] and the values for steels are from [MM].

Table 2.1: Specific Heats of Polymers and Metals

Material	Specific Heat Capacity (J/(kg.K))
ABS	1250-1700
Acetal	1500
PA66	1700
Polycarbonate	1300
Polyethylene	2300
Polypropylene	1900
Polystyrene	1300
PVC (rigid)	800-1200
PVC (plasticized)	1300-2100
Steel (AISI 1020) [MM]	460
Steel (AISI P20) [MM]	460

2.7.2 Thermal Conductivity

Imagine a slab of material of thickness, Δx, for which one side is at a temperature $T + \Delta T$, and the other side is at a *lower* temperature T. Figure (2.6) illustrates this situation.

Figure 2.6: Definition of Thermal Conductivity

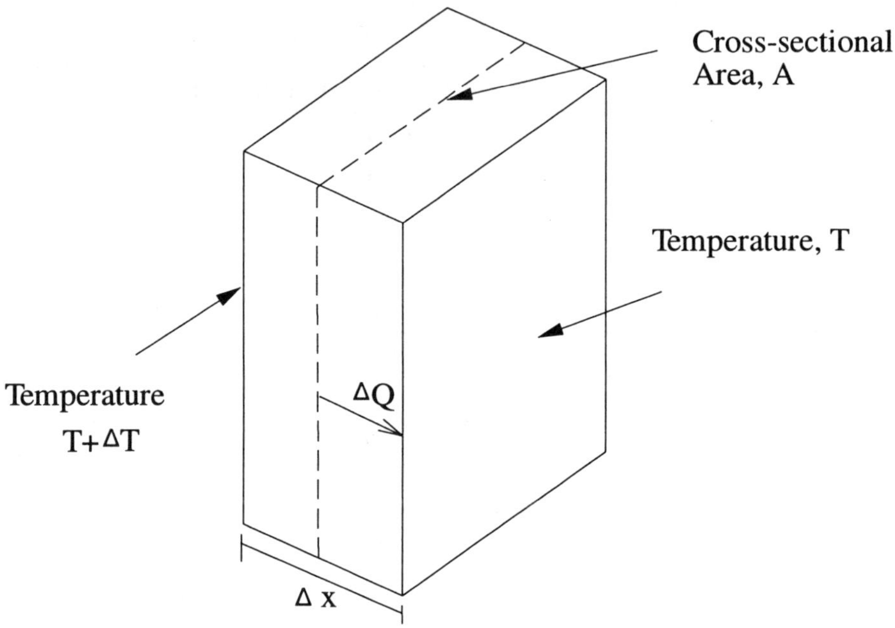

Let ΔQ be the heat flow across the cross-section of area A, in time Δt. The average rate of heat flow across the cross-section is then $\Delta Q/\Delta t$. It is found by experiment, that the average rate of heat flow is proportional to the area and the temperature difference ΔT, and inversely proportional to the distance Δx. That is

$$\frac{\Delta Q}{\Delta t} \propto A\frac{\Delta T}{\Delta x} \tag{2.15}$$

Introducing the *positive* constant of proportionality, k, we have:

$$\frac{\Delta Q}{\Delta t} = -kA\frac{\Delta T}{\Delta x} \tag{2.16}$$

and, in the limit as $\Delta x \to 0$ and $\Delta t \to 0$, we obtain

$$\frac{dQ}{dt} = -kA\frac{dT}{dx} \tag{2.17}$$

The thermal conductivity, k , of the material at temperature, T, is defined by Eq. (2.17) above.

For polymers, the thermal conductivity varies with temperature, degree of crystallinity and level of orientation. It is therefore extremely difficult to measure relevant values of thermal conductivity.

Amorphous materials usually have conductivities in the range from 0.13 J/m.s.K to 0.26 J/m.s.K over the range of processing temperatures. Semi-crystalline materials can have much higher values, the actual value depending on the degree of crystallinity.

2.7.3 No-Flow Temperature

No-flow temperature is defined to be that temperature at which the plastic material effectively ceases to flow. Although not a fundamental physical property of a material, as for example is thermal conductivity, no-flow temperature is useful for modeling the effect of the frozen layer which is deposited on the wall of a mold during the filling phase. In the remainder of this section we briefly consider an alternative method and the use of no-flow temperature in modeling the injection molding process.

Figure (2.7) shows typical temperature and shear rate distributions across the cavity thickness. The corresponding viscosity distribution is shown in Figure (2.8).

Figure 2.7: Shear Rate and Temperature Distribution Across Cavity

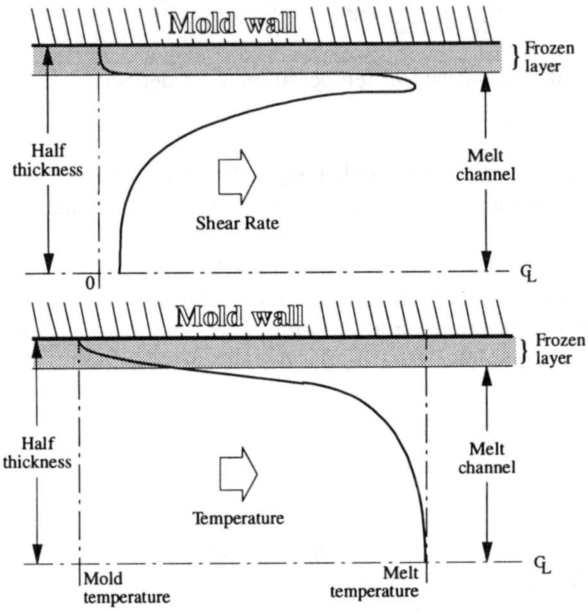

Figure 2.8: Viscosity Distribution Across Cavity

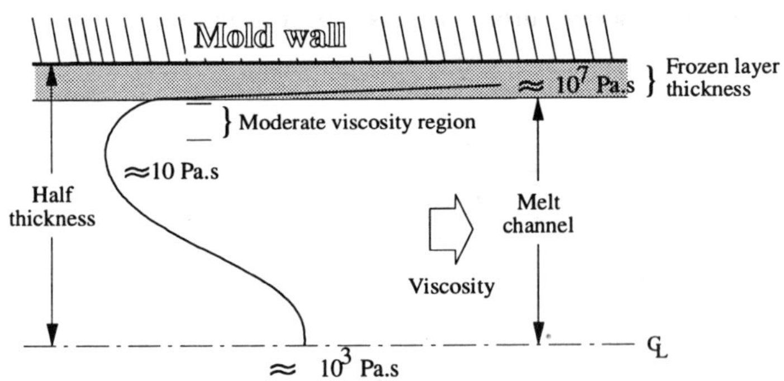

The shape of the viscosity distribution is a consequence of its dependence on both shear rate and temperature. Polymer melts are in general shear thinning. In the absence of temperature effects, the viscosity will decrease with increasing shear rate. On the other hand, if we ignore shear rate, the viscosity tends to decrease with increasing temperature. Both these factors contribute to the shape of the viscosity distribution in Fig. (2.8). Moreover, at low shear rates polymer melts exhibit a Newtonian flow characteristic and this accounts for the lack of change in the viscosity at the cavity centre-line.

The ease with which a melt can be forced through the mold cavity is determined by a quantity called the *fluidity* or *flow conductance*. A careful definition is given in Section (5.3.1). For now we simply define the fluidity S, to be:

$$S = \int_0^h \frac{z^2}{\eta} dz \qquad \textbf{(2.18)}$$

where, h is the half-thickness of the cavity, η is the viscosity and z is the coordinate in the thickness direction measured from the centre line.

Physically, fluidity gives a measure of the ease with which a material flows. High values of fluidity are associated with easy flowing materials — that is, materials offering low resistance to flow. Conversely, a low value of fluidity indicates that the material offers a high resistance to flow.

From Eq. (2.18), it is seen that the greatest contribution to S comes from the region where the viscosity starts to increase rapidly. This is because around the centre line, z is small and so the contribution to S is small, whereas at the wall, while z is large, the viscosity is extremely large and again the contribution to S is small. The most significant region is therefore where the viscosity is not too high but z is not too small. Of course, this region is determined by the thickness of the frozen layer. It follows that the calculation of the fluidity and that of the frozen layer thickness are related.

In injection molding simulation the fluidity is used to calculate the pressure distribution in the cavity. With the pressure distribution known, the velocity of the material can be found. Consequently, calculation of fluidity is a crucial aspect of flow analysis software.

There are two basic approaches to the calculation of fluidity. They are:

- Extrapolate viscosity data so that a viscosity may be defined for the solid material forming the frozen layer and then calculate fluidity S based on a viscosity distribution across the whole section thickness, much in the same way as the description above.
- Use the concept of no-flow temperature to indicate regions where the velocity of the material is effectively zero and so where the viscosity is essentially finite.

We discuss each of these methods in turn.

Extrapolation of Viscosity

In order to calculate fluidity S using the viscosity distribution across the whole section thickness, viscosity data is required at low temperatures, particularly between the important thermal transitions, such as the glass and freezing transitions, and the mold temperature. Measurements in this temperature range are very difficult to make, if they are possible at all. In this region most polymers are essentially solids, so getting them to flow in steady state is almost impossible. This is particularly true of semi-crystalline materials.

In practice, to obtain viscosity values at low temperatures, high temperature viscosity data is extrapolated. While this procedure gives a value for viscosity, the value depends on the temperature dependence of the viscosity model used. The uncertainty introduced by massive extrapolation will introduce uncertainty in the calculation of fluidity and hence pressure predictions.

Figure (2.9) shows actual viscosity versus temperature curves for an amorphous and semi-crystalline material.

Figure 2.9: Errors due to Extrapolation of Viscosity Data

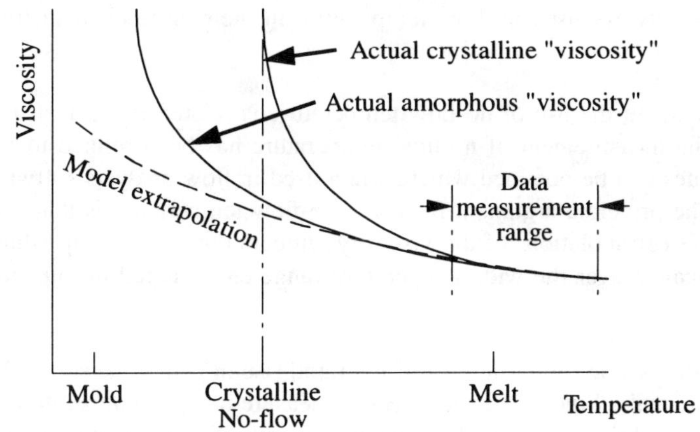

Obviously the change in viscosity is much more rapid for a semi-crystalline material, as it passes through a freezing transition, than for an amorphous material. The diagram shows the typical measurement range of viscosity. This data needs to be represented by a convenient mathematical function thereby defining the relationship between viscosity and temperature. Usually the function fits the measured data well over the measured range. Regardless of the function's ability to fit the data in the measured range however, the function may deviate significantly from the actual behaviour of the material when extrapolated as indicated in the figure. The problem is emphasized by semi-crystalline materials since they exhibit what approaches a step change in viscosity.

Extrapolated viscosity data generally under estimates the true value of viscosity. Using this data in flow analysis software results in an over estimation of fluidity. Physically this means that the material flows more easily than it actually does and so the pressure is under estimated.

Use of No-Flow Temperature

If we turn our attention to the no-flow temperature concept, we note that this in fact tries to remove the extrapolation problem by assigning an infinite viscosity such that excessive extrapolation of viscosity data is not required. Critics of this approach point out that no-flow temperature is not a fundamental physical property of the material. Despite this, the use of no-flow temperature is quite common in flow analysis software [BW], [CH] and [HL]. Moldflow have always used no-flow temperature in their finite element flow analysis software.

Much of the controversy about the use of no-flow temperature is related to its measurement. Over the years, the measurement of no-flow temperature has been refined to the point that repeatable values can be obtained which, when used in flow analysis software give excellent results. The principle argument for using no-flow temperature is that it is not dependent on extreme extrapolation of the viscosity temperature relationship which has little intrinsic significance over the wide temperature range encountered in injection molding.

Use of no-flow temperature is also representative of the rapid rise in resistance shown by a freezing semi-crystalline melt. Since the experiment to measure no-flow temperature is performed under cooling conditions supercooling effects are embodied in the data value. We have observed flow of semi-crystalline materials at temperatures lower than if the material is heated directly to the test temperature. This indicates that supercooling is important. No current viscosity models will show this effect at all.

In addition, since the no-flow temperature falls in the temperature range where the section temperature profile is falling rapidly, small inaccuracies in the measurement do not translate to significant changes in effective section thickness. This means that pressure predictions are relatively insensitive to changes in no-flow temperature.

Sometimes, for amorphous materials, one can generate a viscosity model that shows a moderate rise in viscosity with decreasing temperature. Tests have shown that in this case both the no-flow and extrapolation methods give almost identical velocity profiles, and hence pressures. However this level of agreement does not always occur. In particular, for semi-crystalline materials, which show a rapid viscosity increase due to solidification, the two methods can give quite different results.

For thin wall sections, cooling is so rapid that a significant proportion of the cavity thickness resides at temperatures below the no-flow temperature. Viscosity extrapolation is required over this whole region and will result in incorrect pressure predictions because viscosity in this cold zone will be underestimated. Use of a no-flow temperature will not allow this to occur.

No-flow temperature is also required to indicate short shots — a condition where the flow front is no longer moving due to it, or regions behind it, freezing totally during the filling and/or packing stages. Without a no-flow concept, it is difficult to decide how low flow rate should be before the flow front is considered stationary.

Our conclusion is that no-flow temperature is a most useful concept, and although it is not a fundamental material property, it certainly reflects actual material behaviour. It reduces the uncertainty arising from excessive extrapolation of empirical functions. The characteristic step change in fluidity of semi-crystalline materials is also modeled more effectively. Pressure is not sensitive to small changes in no-flow temperature because temperature gradients are high. We expect that a properly measured no-flow temperature is a far more useful concept than viscosity model extrapolation, particularly for semi-crystalline materials and when analyzing products with thin wall sections.

2.8 Thermodynamic Relationships

Several thermodynamic properties are required for simulation of injection molding. These may be obtained from the PVT data for the material. Before looking at the way this data is handled in Moldflow software, we define the expansivity and compressibility of a material .

2.8.1 Expansivity and Compressibility

An *equation of state* relates the three variables, pressure p, specific volume $\widehat{V}$, and temperature T. For any material, we write the equation of state in the form:

$$f\left(p, \widehat{V}, T\right) = 0 \tag{2.19}$$

Given any two variables, it is possible to determine the third using the equation of state. In particular, we can write

$$\widehat{V} = g(p, T) \tag{2.20}$$

where g is some function. If we graph this data, we would obtain a PVT surface as shown in Fig. (2.10).

Figure 2.10: PVT Diagram

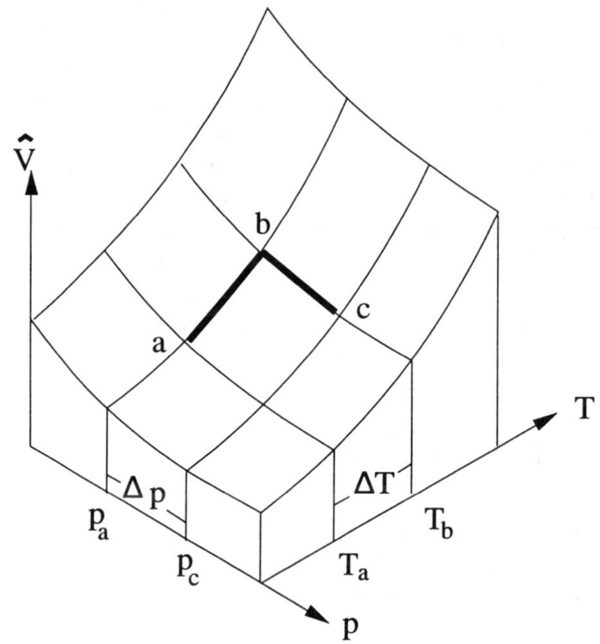

Imagine that the material at temperature T_a, undergoes a change in temperature while being held at constant pressure. What then is the change in volume? This can be answered simply by reference to Fig. (2.10) where this is denoted as a change from point a to point b. A change in temperature from T_a to $T_a + \Delta T$ causes a change in volume equal to $g(p_a, T_a + \Delta T) - g(p_a, T_a)$. The average change in volume over the temperature change is therefore,

$$
\begin{aligned}
\frac{\Delta \widehat{V}}{\Delta T} &= \frac{g(p_a, T_a + \Delta T) - g(p_a, T_a)}{\Delta T} \\
&= \frac{\widehat{V}(p_a, T_a + \Delta T) - \widehat{V}(p_a, Ta)}{\Delta T}
\end{aligned}
\tag{2.21}
$$

In the limit as $\Delta T \rightarrow 0$ we obtain the instantaneous change in volume for the material which we denote by

$$\left(\frac{\partial \widehat{V}}{\partial T}\right)_p \tag{2.22}$$

where the subscript indicates that the pressure is held constant.

The *coefficient of volume expansion* of the material β, is defined as follows:

$$\beta = \frac{1}{\widehat{V}}\left(\frac{\partial \widehat{V}}{\partial T}\right)_p \tag{2.23}$$

The coefficient of volume expansion is also called the *expansivity* of the material. It has units of reciprocal Kelvin (K^{-1}).

Now consider the change in volume due to a change in pressure while keeping the temperature constant. This involves moving from point b to c in Fig. (2.10). The average change in the volume due to a change in pressure is given by

$$\begin{aligned}\frac{\Delta \widehat{V}}{\Delta p} &= \frac{g(p_a + \Delta p, T_b) - g(p_a, T_b)}{\Delta p} \\ &= \frac{\widehat{V}(p_a + \Delta p, T_b) - \widehat{V}(p_a, T_b)}{\Delta p}\end{aligned} \tag{2.24}$$

Once again, letting $\Delta p \to 0$, we obtain the instantaneous change in volume which we denote by

$$\left(\frac{\partial \widehat{V}}{\partial p}\right)_T \tag{2.25}$$

The *isothermal compressibility* κ is defined to be:

$$\kappa = -\frac{1}{\widehat{V}}\left(\frac{\partial \widehat{V}}{\partial p}\right)_T \tag{2.26}$$

The negative sign indicates that the volume decreases with increasing pressure.

Isothermal compressibility has units of one square meter per newton ($m^2\, N^{-1}$).

2.8.2 PVT Data for Flow Analysis

It is essential to have PVT data for the packing phase analysis but it may also be used in the filling phase. The data is obtained experimentally and a curve is fitted using regression analysis. For semi-crystalline materials, PVT data falls into three basic regions: low temperature, transition and high temperature. Figure (2.11) shows a PVT plot for polypropylene. The dotted line indicates the phase transition from melt to solid. Curves to the left of the line are valid for solid material, while curves to the right represent the melt phase.

Figure 2.11: PVT Plot for Semi-Crystalline Material: Polypropylene

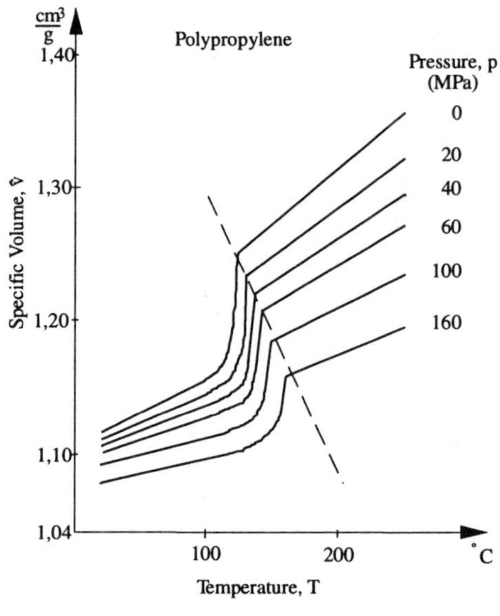

Amorphous materials display no sudden transition from melt to solid. A PVT curve for polystyrene is shown in Fig. (2.12).

Figure 2.12: PVT Plot for Amorphous Material: Polystyrene

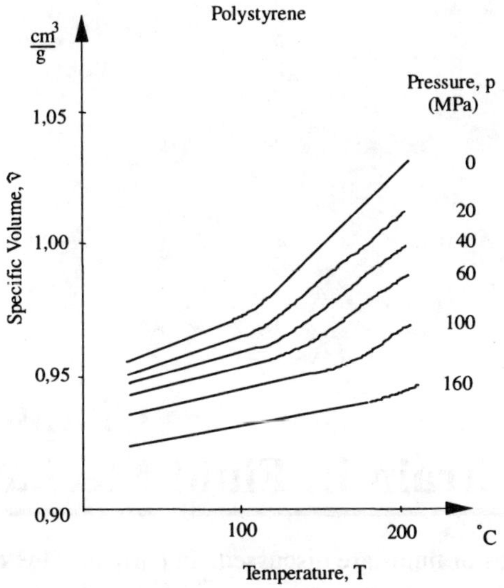

The equations fitted to experimental data to create Figures (2.11) and (2.12) are:

- **Low Temperature Region**

$$\widehat{V} = \frac{a_1}{a_4 + p} + \frac{a_2 T}{a_3 + p} + a_5 e^{a_6 T - a_7 p} \tag{2.27}$$

- **High Temperature Region**

$$\widehat{V} = \frac{a_1}{a_4 + p} + \frac{a_2 T}{a_3 + p} \tag{2.28}$$

- **Transition Region**

$$p = b_1 + b_2 T \tag{2.29}$$

In all of the above, a_i and b_i are constants determined by the regression analysis.

Chapter 3
Stress and Strain in Fluid Mechanics

In this chapter some necessary properties of fluids are discussed. In particular the concept of stress in a fluid is introduced, we also define the rate of strain tensor for a fluid and the concept of a generalized Newtonian fluid.

These ideas are used in subsequent chapters to derive the governing equations for the filling and packing phases of the injection molding process.

3.1 Stress in Fluids

The flow of melt in injection molding involves the deformation of the material due to forces applied by both the molding machine and the mold. Any attempt to determine the flow requires a description of how these forces are transmitted to the melt. The concept of stress allows us to consider the effect of forces on and within the material.

Stress is a measure of the forces transmitted when a force acts on a long body of material. In the following discussion a *body* is simply a mass of material- fluid or solid. There are two basic types of forces: body forces and surface tractions. Body forces are those which act on the elements of mass *within* the body and have the units force per unit volume or force per unit of mass. The most common example is gravity. Surface tractions, as the name suggests, act on the surface of the body and are given in units of force per unit area. An example is the force exerted on an inflated balloon by the gas within. More usually this is called pressure and is directed normal to the wall of the balloon. Surface tractions need not always act normal to the body's surface however. Friction is a case in point. Often frictional forces act tangentially to the surface.

3.1.1 Stress on an External Surface

To define stress in a qualitative way, consider a surface traction $\vec{F}$, acting on the *exterior* surface of a body. Suppose the force acting on a small area, δA, of the surface is $\delta\vec{F}$. The stress vector $\vec{t}$, acting at an external point of a body is defined to be the ratio of the applied force to the area on which the force acts, in the limit as the area goes to zero. That is,

$$\vec{t} = \lim_{\delta A \to 0} \frac{\delta \vec{F}}{\delta A} \tag{3.1}$$

The stress vector $\vec{t}$ completely describes the forces acting at a point on an *external* surface of a body. It may be resolved into components in the x, y and z directions. For example, suppose that the normal to the surface at the point where the stress vector is defined, points in the positive z-direction. Then the stress vector can be resolved into a component in the z-direction, σ_{zz}, and components along the x and y directions, σ_{zx} and σ_{zy} respectively (see Figure (3.13). The component in the z-direction σ_{zz}, is called the *normal stress* while σ_{zx} and σ_{zy} are called *shear stresses*.

Figure 3.13: Resolution of the Stress Vector

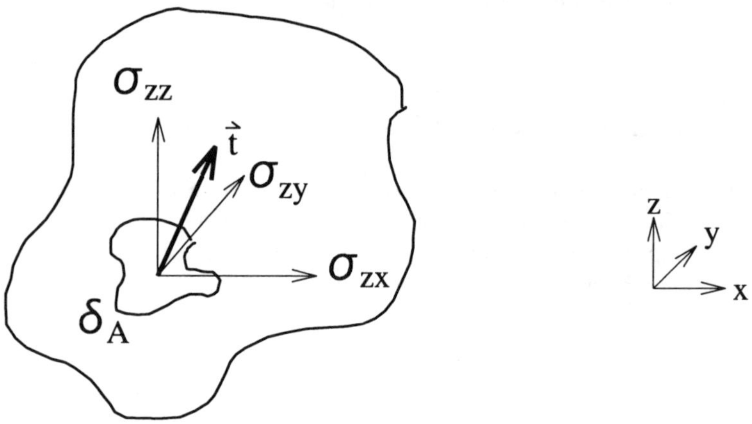

In this notation, the first suffix indicates the normal direction at the point on the surface where the stress vector is defined. The second suffix gives the direction of resolution.

3.1.2 Stress at an Interior Point

While the stress vector defined above defines the stress at a point on the surface, it is not sufficient to define the stress at a point within the body. This is because the stress vector depends on the orientation of the area δA. It can be shown [SH] that if the stresses for three orthogonal surfaces intersecting at a point are known, then the stress at the point can be determined for any surface. To fully describe the state of stress at an interior point, it is therefore necessary to know the stresses on each surface. Each surface will require three stresses, one normal stress and two shear stresses, so a total of nine stresses are required.

Figure (3.14) shows the three surfaces, displaced from the point O for clarity, and the associated stresses acting on them.

Figure 3.14: Orthogonal Surfaces at Interior Point O (shown displaced for clarity)

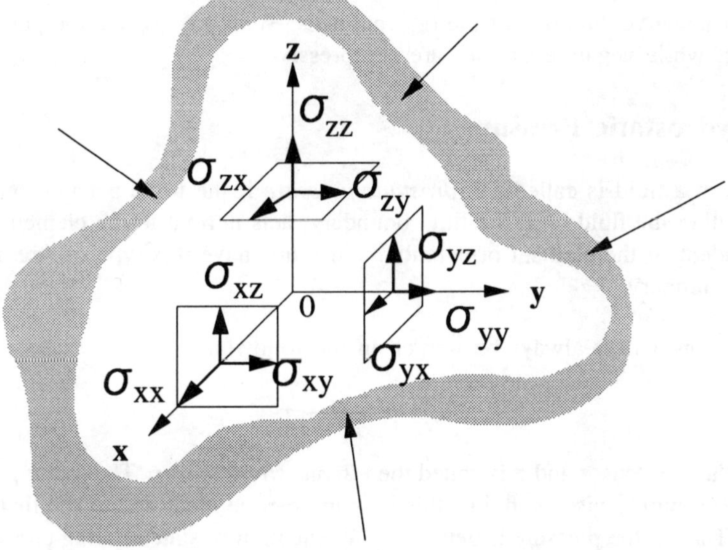

These nine stresses define the state of stress at O. They form the nine components of a second order tensor called the *stress tensor* denoted by $\underline{\sigma}$. The stress tensor is often written as a matrix of its components:

$$\underline{\sigma} = \begin{Bmatrix} \sigma_{xx} & \sigma_{xy} & \sigma_{xz} \\ \sigma_{yx} & \sigma_{yy} & \sigma_{yz} \\ \sigma_{zx} & \sigma_{zy} & \sigma_{zz} \end{Bmatrix} \qquad \textbf{(3.2)}$$

The stress tensor is symmetric. That is, $\sigma_{xz} = \sigma_{zx}$.

Once the stress tensor is known, the stress vector at any surface with unit outward normal $\vec{n}$ is given by:

$$\vec{t} = [\underline{\sigma} \cdot \vec{n}] \qquad \textbf{(3.3)}$$

where the dot indicates a vector-tensor product.

To conclude this section we define a sign convention for stresses. Positive stresses are those for which both the normal to the surface and the direction of the stress point in the positive or negative direction of the relevant axes. With this convention positive stresses are tensile, while negative stresses are compressive.

3.1.3 Hydrostatic Pressure

The stress in a fluid is called a *hydrostatic pressure* if the force per unit on an element of area, within the fluid or at the fluid boundary, acts normal to the element of area and is independent of the element orientation. All fluids have this type of stress state when they are stationary.

The stress tensor may always be written in the following form

$$\underline{\sigma} = -p\underline{I} + \underline{\tau} \quad \textbf{(3.4)}$$

Here $\underline{I}$ is the unit tensor and $\underline{\tau}$ is called the *viscous stress tensor*. The scalar p is called the pressure. For purely viscous fluids, this pressure can be identified as the thermodynamic pressure. That is, the pressure is defined by an equation of state relating pressure, volume and temperature. However, in the case of an incompressible fluid, this is not the case and the pressure is defined as a dynamic variable. We will not discuss this further. Details can be found in [FA].

3.2 Rate of Strain Tensor

General motion of a fluid involves translation, deformation and rotation. The translation of a point in a fluid is defined by its velocity vector $\vec{v}$. The deformation and rotation of the fluid depend on the velocity gradient tensor $\nabla\vec{v}$. We will be concerned with the rate at which the fluid deforms. This is characterized by the rate of strain (or deformation) tensor, $\underline{\dot{\gamma}}$, defined as:

$$\underline{\dot{\gamma}} = \nabla\vec{v} + (\nabla\vec{v})^{\dagger} \quad \textbf{(3.5)}$$

where the dagger indicates the transposed matrix.

Using Cartesian coordinates, $\underline{\dot{\gamma}}$ may be written as a matrix of its components:

$$\underline{\dot{\gamma}} = \begin{bmatrix} 2(\partial v_x/\partial x) & (\partial v_y/\partial x + \partial v_x/\partial y) & (\partial v_x/\partial z + \partial v_z/\partial x) \\ (\partial v_x/\partial y + \partial v_y/\partial x) & 2(\partial v_y/\partial y) & (\partial v_y/\partial z + \partial v_z/\partial y) \\ (\partial v_z/\partial x + \partial v_x/\partial z) & (\partial v_z/\partial y + \partial v_y/\partial z) & 2(\partial v_z/\partial z) \end{bmatrix} \quad \textbf{(3.6)}$$

This tensor describes the rate at which the material changes shape. Like the stress tensor it is also symmetric.

3.3 Compressible and Incompressible Fluids

The principle of conservation of mass leads to the following important equation:

$$\frac{\partial \rho}{\partial t} = -(\nabla \cdot \vec{v}) \quad \textbf{(3.7)}$$

where ρ is the fluid density and $\vec{v}$ the fluid velocity.

This is known as the equation of continuity and is discussed in detail in Section (4.2). An incompressible fluid has constant density and so the continuity equation becomes:

$$\nabla \cdot \vec{v} = 0 \quad \textbf{(3.8)}$$

Hence for an incompressible fluid, $\nabla \cdot \vec{v} = 0$.

3.4 Newtonian and Non-Newtonian Fluids

In order to deal with fluids it is necessary to have a relationship between the viscous stress tensor $\underline{\tau}$ and the rate of strain tensor $\underline{\dot{\gamma}}$. Such a relationship is called a *constitutive equation*.

A class of fluids for which the constitutive equation has a particularly simple form is Newtonian fluids. These have the following constitutive equation [BA]:

$$\underline{\tau} = \mu \underline{\dot{\gamma}} - \left(\frac{2}{3}\mu - \mu_d\right)(\nabla \cdot \vec{v})\underline{I} \quad \textbf{(3.9)}$$

where, μ is the viscosity and μ_d is the *dilatational viscosity* .

The dilatational viscosity is zero for ideal simple gases [BA]. Also, for incompressible flow, we have $\nabla \cdot \vec{v} = 0$ and from Eq. (3.9) we see that in this case the dilatational viscosity has no effect. Since the dilatational viscosity has no effect for these two extremes — a highly compressible gas and an incompressible fluid — we assume its effect to be negligible for a polymer melt and ignore it in what follows.

The constitutive equation for Newtonian fluids then becomes:

$$\begin{aligned} \underline{\tau} &= \mu\underline{\dot{\gamma}} - \frac{2}{3}\mu(\nabla \cdot \vec{v})\underline{I} \\ &= \mu\left(\underline{\dot{\gamma}} - \frac{2}{3}(\nabla \cdot \vec{v})\underline{I}\right) \\ &= \mu\underline{\dot{\gamma}}' \end{aligned} \tag{3.10}$$

where $\underline{\dot{\gamma}}'$ is called the *deviatoric rate of strain tensor* and is defined to be:

$$\underline{\dot{\gamma}}' = \underline{\dot{\gamma}} - \frac{2}{3}(\nabla \cdot \vec{v})\underline{I} \tag{3.11}$$

For an incompressible fluid (see Section 3.3) the deviatoric rate of strain tensor is equal to the rate of strain tensor. In this case, Eq. (3.10) becomes:

$$\underline{\tau} = \mu\underline{\dot{\gamma}} \tag{3.12}$$

Equation (3.12) shows that for an *incompressible* Newtonian fluid, the viscous stress tensor is linearly related to the rate of strain tensor. Note that for such fluids the viscosity is constant at a given temperature. The viscosity may vary with temperature however.

Fluids for which Eq. (3.12) does not hold are called *non-Newtonian.*

Molten polymers are non-Newtonian. Moreover they exhibit *viscoelastic* effects. At this time, Moldflow flow analysis software does not account for viscoelasticity and so we do not consider these effects here.

In order to model the flow of molten polymers, a modification to Eq. (3.10) is made to allow the viscosity to depend on the rate of deformation. Therefore Eq. (3.10) becomes:

$$\underline{\tau} = \eta(\dot{\gamma})\underline{\dot{\gamma}}' \tag{3.13}$$

Where $\eta(\dot{\gamma})$ is called the *viscosity function.* In what follows we will simply call this the viscosity.

Equation (3.13) defines a class of fluids termed *generalized Newtonian fluids*. These types of fluids are useful for modeling polymer flow in situations dominated by shear forces [BA]. Examples include flow in tubes and between flat plates. Clearly this is what we need for modeling the injection molding process.

Substituting Eq.(3.13) in (3.4) we obtain the following relationship, valid for a compressible, generalized Newtonian fluid:

$$\begin{aligned} \underline{\sigma} &= -p\underline{I} + \underline{\tau} \\ &= -p\underline{I} + \eta(\dot{\gamma})\underline{\dot{\gamma}}' \end{aligned} \tag{3.14}$$

For the incompressible case, it is also possible to use the concept of a generalized Newtonian fluid. We simply replace the deviatoric rate of strain tensor with the rate of strain tensor to get:

$$\begin{aligned} \underline{\sigma} &= -p\underline{I} + \underline{\tau} \\ &= -p\underline{I} + \eta(\dot{\gamma})\underline{\dot{\gamma}} \end{aligned} \tag{3.15}$$

Shear Rate and Choice of Viscosity Function

In the previous section the viscosity function, $\eta(\dot{\gamma})$, was introduced to allow the viscosity to depend on the rate of deformation of the fluid. Now it is not sensible to make the viscosity depend directly on the deviatoric rate of strain tensor, $\underline{\dot{\gamma}}'$. This is because $\underline{\dot{\gamma}}'$ depends on the coordinate system in which we work — this certainly isn't true of the viscosity. For this reason, we make the viscosity a function of one of the invariants of the deviatoric rate of strain tensor.

Any second order tensor possesses three invariants. The first, second and third invariants are denoted by I, II and III respectively. For the deviatoric rate of strain tensor they are defined as follows:

$$\begin{aligned} I_{\dot{\gamma}'} &= \sum_i \dot{\gamma}'_{ii} \\ II_{\dot{\gamma}'} &= \sum_i \sum_j \dot{\gamma}'_{ij}\dot{\gamma}'_{ji} \\ III_{\dot{\gamma}'} &= \sum_i \sum_j \sum_k \dot{\gamma}'_{ij}\dot{\gamma}'_{jk}\dot{\gamma}'_{ki} \end{aligned} \tag{3.16}$$

Using equations (3.6) and (3.11) we find that the first invariant is identically zero:

$$I_{\dot{\gamma}'} = 2\left(\frac{\partial v_x}{\partial x} + \frac{\partial v_y}{\partial y} + \frac{\partial v_z}{\partial z}\right) - 2(\nabla \cdot \vec{v}) \\ = 0 \tag{3.17}$$

and is therefore not useful for use in the viscosity function.

In addition, it can be shown [BA] that the third invariant is zero for shearing type flow. Since this type of flow is frequently encountered in injection molding, we allow the viscosity to depend on the second invariant $II_{\dot{\gamma}'}$.

For symmetric tensors, the second invariant is related to the magnitude of the tensor. The double dot (or scalar product) of two tensors, $\underline{a}$, $\underline{b}$ is defined as follows:

$$\underline{a} : \underline{b} = \sum_i \sum_j a_{ij} b_{ji} \tag{3.18}$$

The magnitude of a tensor $\underline{\dot{\gamma}}$ is denoted by $\dot{\gamma}$ and defined as follows

$$\dot{\gamma} = \sqrt{\frac{1}{2} \underline{\dot{\gamma}} : \underline{\dot{\gamma}}^{\dagger}} \tag{3.19}$$

Both deviatoric and the usual strain rate tensors are symmetric and so $\underline{\dot{\gamma}}'_{ij} = \underline{\dot{\gamma}}'_{ji}$. Using this fact and the definitions of the second invariant and double dot product we obtain

$$\begin{aligned} II_{\dot{\gamma}'} &= \sum_i \sum_j \dot{\gamma}'_{ij} \dot{\gamma}'_{ji} \\ &= \sum_i \sum_j \dot{\gamma}'_{ij} \dot{\gamma}'_{ij} \\ &= \underline{\dot{\gamma}}' : \underline{\dot{\gamma}}'^{\dagger} \\ &= 2\dot{\gamma}'^2 \end{aligned} \tag{3.20}$$

The quantity $\dot{\gamma}'$ is defined to be the *magnitude* of the deviatoric rate of strain tensor. More usually, $\dot{\gamma}'$ is known as the *shear rate*. Rearranging, we obtain

$$\begin{aligned} \dot{\gamma}' &= \sqrt{\frac{1}{2} II_{\dot{\gamma}'}} \\ &= \sqrt{\frac{1}{2} \underline{\dot{\gamma}}' : \underline{\dot{\gamma}}'} \end{aligned} \tag{3.21}$$

We have defined the shear rate in terms of the magnitude of the deviatoric rate of strain tensor. Exactly the same relationships for the rate of strain tensor may be used in the case of an incompressible fluid. That is,

$$\begin{aligned}\dot{\gamma} &= \sqrt{\frac{1}{2} II_{\dot{\gamma}}} \\ &= \sqrt{\frac{1}{2} \underline{\dot{\gamma}} : \underline{\dot{\gamma}}}\end{aligned} \tag{3.22}$$

It is convenient to allow the viscosity function to depend on the shear rate as defined above. Some examples of viscosity functions were given in Chapter 2.

To conclude this Chapter we consider the form of the shear rate in a Cartesian coordinate system for a simple shear flow. Such a flow is of interest as it describes the flow of melt in the cavity of an injection mold.

Adopting a Cartesian coordinate system, for a steady shearing flow in the x-y plane we find from Eq. (3.6) that the rate of strain tensor has the matrix form:

$$\underline{\dot{\gamma}} = \begin{bmatrix} 0 & 0 & \partial v_x/\partial z \\ 0 & 0 & \partial v_y/\partial z \\ \partial v_x/\partial z & \partial v_y/\partial z & 0 \end{bmatrix} \tag{3.23}$$

Since the flow is assumed steady, $\nabla \cdot \vec{v} = 0$ and the deviatoric rate of strain tensor and rate of strain tensor are equal. That is,

$$\underline{\dot{\gamma}}' = \underline{\dot{\gamma}} = \begin{bmatrix} 0 & 0 & \partial v_x/\partial z \\ 0 & 0 & \partial v_y/\partial z \\ \partial v_x/\partial z & \partial v_y/\partial z & 0 \end{bmatrix} \tag{3.24}$$

The second invariant therefore has the value:

$$\begin{aligned}II_{\dot{\gamma}} &= \sum_i \sum_j \dot{\gamma}_{ij} \dot{\gamma}_{ji} \\ &= 2\dot{\gamma}_{xz}\dot{\gamma}_{zx} + 2\dot{\gamma}_{zy}\dot{\gamma}_{yz} \\ &= 2\left(\frac{\partial v_x}{\partial z}\right)^2 + 2\left(\frac{\partial v_y}{\partial z}\right)^2\end{aligned} \tag{3.25}$$

As we saw earlier, the shear rate is related to the second invariant as follows:

$$\dot{\gamma}' = \sqrt{\frac{1}{2} II_{\dot{\gamma}'}} \tag{3.26}$$

For steady shearing flow in the x–y plane the shear rate is, from Eq. (3.25), given by

$$\begin{aligned} \dot{\gamma}' &= \sqrt{\frac{1}{2} II_{\dot{\gamma}'}} \\ &= \sqrt{\left(\frac{\partial v_x}{\partial z}\right)^2 + \left(\frac{\partial v_y}{\partial z}\right)^2} \end{aligned} \tag{3.27}$$

Chapter 4
Governing Equations for Fluid Flow

In this chapter, the equations governing the flow of a compressible, viscous fluid are derived. These equations are applicable to the flow of a polymer melt and are obtained using the principles of conservation of mass, momentum and energy.

After deriving the governing equations, boundary conditions that are appropriate for the modeling of the injection molding process are described.

4.1 Mathematical Preliminaries

In deriving the governing equations for a fluid (in particular a polymer melt), we assume that the fluid is a continuum. This means that we ignore the molecular structure of the material and assume that it is possible to define physical variables such as velocity and density at a point in the fluid. Moreover, we assume that the value of these variables change smoothly so that differentiation is permissible. While neglect of the molecular structure of fluids is a big assumption, it turns out that the continuum hypothesis allows a very accurate description of fluid motion and allows us to use standard mathematical techniques.

In this section we introduce some mathematical concepts which will be used in later Chapters. They are the concept of material derivative, Reynold's transport theorem and the Divergence Theorem of Gauss. The latter is also known as the Gauss-Ostragradskii Theorem.

4.1.1 The Material Derivative

Some fluid properties are functions of both position and time. When analyzing fluid motion it is necessary to take derivatives, with respect to time, of these properties. For example, the density of a fluid particle may depend on both position and time. In Cartesian coordinates we can write,

$$\rho = \rho(x, y, z, t) \tag{4.1}$$

where x, y and z are the coordinates of the fluid particle.

Using the chain rule for differentiation, we find the derivative of the density with respect to time is given by

$$\begin{aligned} \frac{d\rho}{dt} &= \frac{\partial \rho}{\partial x}\frac{\partial x}{\partial t} + \frac{\partial \rho}{\partial y}\frac{\partial y}{\partial t} + \frac{\partial \rho}{\partial z}\frac{\partial z}{\partial t} + \frac{\partial \rho}{\partial t} \\ &= \frac{\partial \rho}{\partial x}\dot{x}(t) + \frac{\partial \rho}{\partial y}\dot{y}(t) + \frac{\partial \rho}{\partial z}\dot{z}(t) + \frac{\partial \rho}{\partial t} \\ &= \vec{v}.\nabla\rho + \frac{\partial \rho}{\partial t} \end{aligned} \tag{4.2}$$

where $\vec{v} = (\dot{x}, \dot{y}, \dot{z})$.

More generally for any scalar function of position and time, $f(x, y, z, t)$, we have

$$\frac{df}{dt} = (\vec{v} \cdot \nabla f) + \frac{\partial f}{\partial t} \tag{4.3}$$

Frequently the notation

$$\frac{D}{Dt} = \vec{v} \cdot \nabla + \frac{\partial}{\partial t} \tag{4.4}$$

is used to denote such differentiation. We call Df/Dt the material derivative of f (also called the convected or substantive derivative). Physically, Df/Dt takes into account both the motion of the fluid and the changing position of the fluid particle with time.

It is possible to take a material derivative of scalar, vector or tensor quantities that vary with position and time. For example, the material derivative of the velocity $\vec{v}$, of a fluid particle, is given by

$$\frac{D\vec{v}}{Dt} = [\vec{v} \cdot \nabla\vec{v}] + \frac{\partial \vec{v}}{\partial t} \tag{4.5}$$

The material derivative of a tensor quantity $\underline{B}$, is given by

$$\frac{D\underline{B}}{Dt} = \{\vec{v} \cdot \nabla\underline{B}\} + \frac{\partial \underline{B}}{\partial t} \tag{4.6}$$

4.1.2 Gauss's Divergence Theorem

This important result relates volume and surface integrals. It may be stated as follows:

$$\int_V \left(\nabla \cdot \vec{f}\right) dV = \int_S \left(\vec{f} \cdot \vec{n}\right) dS \tag{4.7}$$

where V is a region in space, S is the closed boundary of V, $\vec{n}$ is a unit outward normal to S and $\vec{f}$ is any smooth vector function. There are technical requirements on the region V but we will not worry about them here. Details may be found in [MT].

We will also use a similar result that is valid in lower dimensions. This may be called the divergence theorem in the plane and is stated as:

$$\int_A (\nabla \cdot f) dA = \int_S \left(\vec{f} \cdot \vec{n}\right) dS \tag{4.8}$$

where A is an area in the plane, S is the closed boundary of A, $\vec{n}$ is a unit outward normal to S and $\vec{f}$ is any smooth vector function. Again we ignore the technical requirements [MT] on A and S.

4.1.3 Reynold's Transport Theorem

To derive the governing equations, we will need to consider integrals of any function of position and time $f(\vec{x}, t)$ over a volume of fluid. This volume will move with the fluid but consists of the same fluid particles. Such a volume is called a *material volume* and is denoted $V(t)$.

The expression

$$F(t) = \int_{V(t)} f(\vec{x}, t) dV \tag{4.9}$$

defines a function of t. Reynold's Transport Theorem [AR] tells us how to calculate the derivative of $F(t)$ with respect to time. Note that because the volume $V(t)$ varies with time and moves with the fluid, it is not possible to take the differentiation through the integral sign. Moreover it is necessary to use the material derivative of $F(t)$.

The theorem may be written as:

$$\frac{d}{dt} \int_{V(t)} f(\vec{x}, t) dV = \int_{V(t)} \left\{ \frac{\partial f}{\partial t} + (\nabla \cdot f\vec{v}) \right\} dV \tag{4.10}$$

where $\vec{v}$ is the velocity of a fluid particle.

Applying the divergence theorem to the RHS of Eq. (4.10) we obtain the following equivalent form of the transport theorem:

$$\frac{d}{dt} \int_{V(t)} f(\vec{x}, t) dV = \int_{V(t)} \frac{\partial f}{\partial t} dV + \int_{S(t)} f(\vec{v} \cdot \vec{n}) dS \tag{4.11}$$

where $S(t)$ is the surface of $V(t)$ and $\vec{n}$ is the outwardly directed unit normal to $S(t)$.

Physically, equation (4.11) states that the rate of change of the integral of $f(\vec{x}, t)$ is equal to the integral of the rate of change of $f(\vec{x}, t)$ over a fixed region plus the resultant flow of $f(\vec{x}, t)$ across the surface $S(t)$. The result holds for any scalar, vector or tensor function $f(\vec{x}, t)$.

4.2 Conservation of Mass

If $V(t)$ is a material volume of fluid, in which there are no sources or sinks, then conservation of mass means that the mass contained in $V(t)$ does not change.

The mass m within $V(t)$ is given by

$$m = \int\limits_{V(t)} \rho(\vec{x}, t) dV \tag{4.12}$$

where $\rho(\vec{x}, t)$ is the density of the fluid at the point $\vec{x}$ and time t.

Conservation of mass implies that the rate of change of m with respect to t is zero. That is,

$$\begin{aligned} 0 &= \frac{dm}{dt} \\ &= \frac{d}{dt} \int\limits_{V(t)} \rho(\vec{x}, t) dV \end{aligned} \tag{4.13}$$

Note that the material derivative is used because the volume $V(t)$ moves with the fluid.

Applying Reynold's Transport Theorem (see Section 4.1.3) to the RHS we obtain:

$$0 = \int\limits_{V(t)} \left\{ \frac{\partial \rho}{\partial t} + (\nabla \cdot \rho \vec{v}) \right\} dV \tag{4.14}$$

Since the volume, $V(t)$, is arbitrary, the integrand must be identically zero. That is,

$$\frac{\partial \rho}{\partial t} + (\nabla \cdot \rho \vec{v}) = 0 \tag{4.15}$$

This equation is called the *continuity equation for a compressible fluid.*

Now, for an incompressible fluid (see Section 3.3), the density is constant and so Eq. (4.15) becomes:

$$(\nabla \cdot \vec{v}) = 0 \tag{4.16}$$

Sometimes it is useful to express the continuity equation in terms of the material derivative. Expanding out Eq. (4.15) we get:

$$\frac{\partial \rho}{\partial t} + \rho(\nabla \cdot \vec{v}) + (\vec{v} \cdot \nabla \rho) = 0 \tag{4.17}$$

Using the definition of material derivative, (see Section 4.1.1), we may write this in the following form:

$$\frac{D\rho}{Dt} = -\rho(\nabla \cdot \vec{v}) \tag{4.18}$$

4.3 Conservation of Momentum

Conservation of momentum requires that the time rate of change of fluid particle momentum in a material volume, $V(t)$, be equal to the sum of external forces acting on $V(t)$. That is,

$$\frac{d}{dt}\int\limits_{V(t)} \rho\vec{v}dV = \sum \vec{F}_{ext} \tag{4.19}$$

The external forces acting on V include both body forces (due to gravity) and tractions. The total body force, $\vec{F}_b$, is given by

$$\vec{F}_b = \int\limits_{V(t)} \rho\vec{g}\,dV \tag{4.20}$$

where $\vec{g}$ is the total body force per unit mass. Usually $\vec{g}$ is due to gravitational effects.

The traction force acting on an element dS, of the boundary surface of $V(t)$ is given by $\vec{t}dS$ where $\vec{t}$ is the stress vector (see Section 3.1.1). The total traction force, $\vec{F}_t$, is therefore given by

$$\begin{aligned} \vec{F}_t &= \int\limits_{S(t)} \vec{t}\,dS \\ &= \int\limits_{S(t)} \underline{\sigma} \cdot \vec{n}\,dS \end{aligned} \tag{4.21}$$

where $S(t)$ is the boundary surface of $V(t)$, $\underline{\sigma}$ is the stress tensor (see Section 3.1.2) and $\vec{n}$ is the unit outward normal to $S(t)$. Applying the Divergence Theorem (Section 4.1.2), we obtain

$$\vec{F}_t = \int_{V(t)} [\nabla \cdot \underline{\sigma}] dV \tag{4.22}$$

From equations (4.20) and (4.22) the total external force is given by:

$$\begin{aligned} \sum \vec{F}_{ext} &= \vec{F}_b + \vec{F}_t \\ &= \int_{V(t)} \rho \vec{g} dV + \int_{V(t)} [\nabla \cdot \underline{\sigma}] dV \end{aligned} \tag{4.23}$$

Substituting this result in (4.19) gives

$$\frac{d}{dt} \int_V \rho \vec{v} dV = \int_V \rho \vec{g} dV + \int_V [\nabla \cdot \underline{\sigma}] dV \tag{4.24}$$

Applying the Transport Theorem (see Section 4.1.3)) to the LHS of the above equation we obtain,

$$\int_V \left\{ \frac{\partial}{\partial t} (\rho \vec{v}) + [\nabla \cdot \{\rho \vec{v} \vec{v}\}] \right\} dV = \int_V \rho \vec{g} dV + \int_V [\nabla \cdot \underline{\sigma}] dV \tag{4.25}$$

Since $V(t)$ is arbitrary, we have after rearrangement,

$$\frac{\partial}{\partial t} (\rho \vec{v}) = \rho \vec{g} + [\nabla \cdot \underline{\sigma}] - [\nabla \cdot (\rho \vec{v} \vec{v})] \tag{4.26}$$

This equation is called *the momentum equation.*

4.4 Conservation of Energy

The total energy of fluid in a material volume $V(t)$ is given by the sum of its kinetic and internal energies. If $\hat{U}$ is the internal energy per unit volume, the total energy is therefore given by:

$$\int_{V(t)} \rho \left(\frac{1}{2} v^2 + \hat{U} \right) dV = \int_{V(t)} \frac{1}{2} \rho v^2 dV + \int_{V(t)} \rho \hat{U} dV \tag{4.27}$$

The first law of thermodynamics states that the increase in total energy of a material volume $V(t)$, with boundary $S(t)$ is equal to the work done on the volume less the heat lost through $S(t)$.

The rate of work done on the material volume by external forces is due to both traction and body forces.

The rate of work done by traction forces is given by

$$\int_{S(t)} (\vec{v} \cdot \vec{t}) dS = \int_{S(t)} (\vec{v} \cdot [\underline{\sigma} \cdot n]) dS \qquad \textbf{(4.28)}$$

where we have used the definition of the stress vector $\vec{t}$. Using the Divergence Theorem, the RHS of the above can be written as a volume integral. That is,

$$\begin{aligned} \int_{S(t)} (\vec{v} \cdot \vec{t}) dS &= \int_{S(t)} (\vec{v} \cdot [\underline{\sigma} \cdot n]) dS \\ &= \int_{V(t)} (\nabla \cdot [\vec{v} \cdot \underline{\sigma}]) dV \end{aligned} \qquad \textbf{(4.29)}$$

We will only consider body forces due to gravity. Denoting the acceleration due to gravity by $\vec{g}$, the rate of work done on the material volume by gravitational forces is given by

$$\int_{V(t)} \rho (\vec{g} \cdot \vec{v}) dV \qquad \textbf{(4.30)}$$

The total rate of work done on the material volume due to both traction and gravitational forces is, therefore the sum of equations (4.29) and (4.30).

Consider now the heat flux across the surface $S(t)$. Let $\vec{q}$ denote the heat flux at a point on $S(t)$. The rate of heat loss from $S(t)$ is given by

$$-\int_{S(t)} (\vec{q} \cdot \vec{n}) dS = -\int_{V(t)} (\nabla \cdot \vec{q}) dV \qquad \textbf{(4.31)}$$

where again, we have used the Divergence Theorem to transform the surface integral to a volume integral.

Equating the sum of equations (4.29), (4.30) and (4.31) to the rate of increase in total energy gives, using equation (4.27)

$$\underbrace{\frac{d}{dt}\int_V \frac{1}{2}\rho v^2 dV}_{(i)} + \underbrace{\frac{d}{dt}\int_V \rho\widehat{U} dV}_{(ii)} = \underbrace{\int_V (\nabla\cdot[\vec{v}\cdot\underline{\sigma}])dV}_{(iii)} + \underbrace{\int_V \rho(\vec{g}\cdot\vec{v})dV}_{(iv)} - \underbrace{\int_V (\nabla\cdot\vec{q})dV}_{(v)} \quad \textbf{(4.32)}$$

At this point it is worth considering the physical meaning of equation (4.32). For convenience each expression in the equation has been labelled. The physical significance of the terms are:

i. is the rate of change of kinetic energy of the material volume.

ii. is the rate of change of internal energy of the material volume.

iii. is the rate of work done on the fluid within the material volume by viscous forces.

iv. is the rate of work done on the fluid within the material volume by gravitational forces.

v. is the heat loss from the fluid within the material volume due to conduction across $S(t)$.

It can be shown [AR] that the rate of change of kinetic energy is given by

$$\frac{d}{dt}\int_{V(t)} \frac{1}{2}\rho v^2 dV = \int_{V(t)} \rho(\vec{g}\cdot\vec{v})dV - \int_{V(t)} (\underline{\sigma} : \{\nabla\vec{v}\})dV + \int_{V(t)} (\nabla\cdot[\vec{v}\cdot\underline{\sigma}])dV \quad \textbf{(4.33)}$$

Substituting this result in equation (4.32) we get

$$\frac{d}{dt}\int_{V(t)} \rho\widehat{U} dV = \int_{V(t)} (\underline{\sigma} : \{\nabla\vec{v}\})dV - \int_{V(t)} (\nabla\cdot\vec{q})dV \quad \textbf{(4.34)}$$

Applying the transport theorem to the LHS we obtain,

$$\int_{V(t)} \left(\frac{\partial}{\partial t}\left(\rho\widehat{U}\right) + \left(\nabla\cdot\rho\widehat{U}\vec{v}\right)\right)dV = \int_{V(t)} (\underline{\sigma} : \{\nabla\vec{v}\})dV - \int_{V(t)} (\nabla\cdot\vec{q})dV \quad \textbf{(4.35)}$$

Because the region $V(t)$ is arbitrary, we must have:

$$\frac{\partial}{\partial t}\left(\rho\widehat{U}\right) + \left(\nabla \cdot \rho\widehat{U}\vec{v}\right) = (\underline{\sigma} : \{\nabla\vec{v}\}) - (\nabla \cdot \vec{q}) \tag{4.36}$$

Expanding the LHS and using the definition of material derivative (see Section 4.1.1), we have:

$$\begin{aligned}
\frac{\partial}{\partial t}\left(\rho\widehat{U}\right) + \left(\nabla \cdot \rho\widehat{U}\vec{v}\right) &= \frac{\partial}{\partial t}\left(\rho\widehat{U}\right) + \rho\widehat{U}(\nabla \cdot \vec{v}) + \left(\vec{v} \cdot \nabla\rho\widehat{U}\right) \\
&= \frac{D}{Dt}\left(\rho\widehat{U}\right) + \rho\widehat{U}(\nabla \cdot \vec{v}) \\
&= \rho\frac{D\widehat{U}}{Dt} + \widehat{U}\frac{D\rho}{Dt} + \rho\widehat{U}(\nabla \cdot \vec{v}) \\
&= \rho\frac{D\widehat{U}}{Dt}
\end{aligned} \tag{4.37}$$

where, in the last line, we have used the equation of continuity, Eq. (4.18).

Using Eq. (4.37), we obtain the following form of Eq. (4.36):

$$\rho\frac{D\widehat{U}}{Dt} = (\underline{\sigma} : \{\nabla\vec{v}\}) - (\nabla \cdot \vec{q}) \tag{4.38}$$

4.4.1 Relating Specific Energy to Temperature

Usually it is more convenient to write equation (4.38) in terms of temperature T, rather than internal energy $\hat{U}$. In this section we derive such a relation.

It should be noted that the material in this section assumes some knowledge of thermodynamics. Although this section is not required for an understanding of the remainder of this book, it is included for completeness. Readers unfamiliar with thermodynamical concepts should move to the next section.

Let

$$\widehat{H} = \widehat{U} + p\widehat{V} \tag{4.39}$$

where p is pressure and $\hat{V}$ is specific volume, denote the specific enthalpy. A combination of the first and second law of thermodynamics may be written in the following form [SS]:

$$d\widehat{S} = \frac{1}{T}\left(d\widehat{H} - \widehat{V}\,dp\right) \tag{4.40}$$

where $\hat{S}$ is the specific entropy. We will suppose that $\hat{H}$ is a function of both pressure and temperature, that is $\hat{H} = \hat{H}(p,T)$. Then

$$d\widehat{H} = \left(\frac{\partial \widehat{H}}{\partial p}\right)_T dp + \left(\frac{\partial \widehat{H}}{\partial T}\right)_p dT \tag{4.41}$$

Substituting this expression in (4.40), we get

$$d\widehat{S} = \frac{1}{T}\left[\left(\frac{\partial \widehat{H}}{\partial p}\right)_T - \widehat{V}\right] dp + \frac{1}{T}\left(\frac{\partial \widehat{H}}{\partial T}\right)_p dT \tag{4.42}$$

But $\hat{S}$ is also a function of p and T, therefore

$$d\widehat{S} = \left(\frac{\partial \widehat{S}}{\partial p}\right)_T dp + \left(\frac{\partial \widehat{S}}{\partial T}\right)_p dT \tag{4.43}$$

Equating coefficients of (4.42) and (4.43) we obtain

$$\left(\frac{\partial \widehat{S}}{\partial p}\right)_T = \frac{1}{T}\left[\left(\frac{\partial \widehat{H}}{\partial p}\right)_T - \widehat{V}\right] \tag{4.44}$$

and

$$\left(\frac{\partial \widehat{S}}{\partial T}\right)_p = \frac{1}{T}\left(\frac{\partial \widehat{H}}{\partial T}\right)_p \tag{4.45}$$

We now differentiate equation (4.44) with respect to T to get,

$$\begin{aligned}\frac{\partial^2 \widehat{S}}{\partial T \partial p} &= \left[\frac{\partial}{\partial T}\left(\frac{\partial \widehat{S}}{\partial p}\right)_T\right]_p \\ &= -\frac{1}{T^2}\left[\left(\frac{\partial \widehat{H}}{\partial p}\right)_T - \widehat{V}\right] + \frac{1}{T}\left(\left[\frac{\partial}{\partial T}\left(\frac{\partial \widehat{H}}{\partial p}\right)_T\right]_p - \left(\frac{\partial \widehat{V}}{\partial T}\right)_p\right) \\ &= -\frac{1}{T^2}\left[\left(\frac{\partial \widehat{H}}{\partial p}\right)_T - \widehat{V}\right] + \frac{1}{T}\left(\frac{\partial^2 \widehat{H}}{\partial T \partial p} - \left(\frac{\partial \widehat{V}}{\partial T}\right)_p\right)\end{aligned} \tag{4.46}$$

Now differentiate equation (4.45) with respect to p to get,

$$\begin{aligned}\frac{\partial^2 \widehat{S}}{\partial p \partial T} &= \left[\frac{\partial}{\partial p}\left(\frac{\partial \widehat{S}}{\partial T}\right)_p\right]_T \\ &= \frac{1}{T}\left[\frac{\partial}{\partial p}\left(\frac{\partial \widehat{H}}{\partial T}\right)_p\right]_T \\ &= \frac{1}{T}\frac{\partial^2 \widehat{H}}{\partial T \partial p}\end{aligned} \tag{4.47}$$

Because

$$\frac{\partial^2 \widehat{S}}{\partial T \partial p} = \frac{\partial^2 \widehat{S}}{\partial p\, \partial T} \tag{4.48}$$

we can equate equations (4.46) and (4.47) and so obtain,

$$\left(\frac{\partial \widehat{H}}{\partial p}\right)_T = \widehat{V} - T\left(\frac{\partial \widehat{V}}{\partial T}\right)_p \tag{4.49}$$

Substituting this result in equation (4.41) we get

$$d\widehat{H} = \left(\widehat{V} - T\left(\frac{\partial \widehat{V}}{\partial T}\right)_p\right) dp + \left(\frac{\partial \widehat{H}}{\partial T}\right)_p dT \tag{4.50}$$

This equation can be put into a simpler form by noting that the specific heat of a material under constant pressure c_p, and the coefficient of volume expansion β are related to the partial derivatives as follows:

$$c_p = \left(\frac{\partial \widehat{H}}{\partial T}\right)_p, \qquad \beta = \frac{1}{\widehat{V}}\left(\frac{\partial \widehat{V}}{\partial T}\right)_p \tag{4.51}$$

Hence Eq. (4.50) may be written as

$$d\widehat{H} = (1 - \beta T)\widehat{V}\,dp + c_p dT \tag{4.52}$$

Using equation (4.39) we find

$$d\widehat{H} = d\widehat{U} + \widehat{V}dp + pd\widehat{V} \tag{4.53}$$

Substituting for $d\hat{H}$, and rearranging for $d\hat{U}$ we obtain

$$\begin{aligned} d\widehat{U} &= (1 - \beta T)\widehat{V}dp + c_p dT - pd\widehat{V} - \widehat{V}dp \\ &= c_p dT - \beta T\widehat{V}dp - pd\widehat{V} \end{aligned} \tag{4.54}$$

The derivative of the specific energy with respect to time is therefore given by:

$$\frac{d\widehat{U}}{dt} = c_p\frac{dT}{dt} - \beta T\widehat{V}\frac{dp}{dt} - p\frac{d\widehat{V}}{dt} \tag{4.55}$$

It follows that in terms of material derivatives, we have

$$\frac{D\widehat{U}}{Dt} = c_p\frac{DT}{Dt} - \beta T\widehat{V}\frac{Dp}{Dt} - p\frac{D\widehat{V}}{Dt} \tag{4.56}$$

Now consider the term $D\hat{V}/Dt$. Because $\hat{V} = 1/\rho$ we have,

$$\begin{aligned}
\frac{D\widehat{V}}{Dt} &= \frac{\partial \widehat{V}}{\partial t} + \vec{v} \cdot \nabla \widehat{V} \\
&= \frac{\partial}{\partial t}\left(\frac{1}{\rho}\right) + \vec{v} \cdot \nabla \left(\frac{1}{\rho}\right) \\
&= \frac{\partial}{\partial \rho}\left(\frac{1}{\rho}\right)\frac{\partial \rho}{\partial t} + \vec{v} \cdot \left[\frac{\partial}{\partial \rho}\left(\frac{1}{\rho}\right)\nabla \rho\right] \\
&= -\frac{1}{\rho^2}\left(\frac{\partial \rho}{\partial t} + \vec{v} \cdot \nabla \rho\right) \\
&= -\frac{1}{\rho^2}\frac{D\rho}{Dt} \\
&= \frac{1}{\rho}\nabla \cdot \vec{v}
\end{aligned} \tag{4.57}$$

where, in the last equation we have used Eq. (4.18).

Substitution of the above in Eq. (4.56) and multiplying through by the density gives

$$\rho\frac{D\widehat{U}}{Dt} = \rho c_p \frac{DT}{Dt} - \beta T \frac{Dp}{Dt} - p\nabla \cdot \vec{v} \tag{4.58}$$

This expression relates the rate of change of specific energy to rate of change of temperature. In the next section this result is used to express the energy equation in terms of temperature.

4.4.2 The Energy Equation in Terms of Temperature

In the preceding section it was shown that

$$\rho\frac{D\widehat{U}}{Dt} = \rho c_p \frac{DT}{Dt} - \beta T \frac{Dp}{Dt} - p\nabla \cdot \vec{v} \tag{4.59}$$

This result can be used to express the energy equation derived earlier in terms of temperature. The energy equation, from Eq. (4.38) is

$$\rho\frac{D\widehat{U}}{Dt} = (\underline{\sigma} : \{\nabla\vec{v}\}) - (\nabla \cdot \vec{q}) \tag{4.60}$$

Substitution of Eq. (4.59) in the above expression gives

$$\rho c_p \frac{DT}{Dt} = \beta T \frac{Dp}{Dt} + p\nabla \cdot \vec{v} + (\underline{\sigma} : \{\nabla \vec{v}\}) - (\nabla \cdot \vec{q}) \qquad \textbf{(4.61)}$$

Expanding the material derivatives we obtain:

$$\rho c_p \left(\frac{\partial T}{\partial t} + \vec{v} \cdot \nabla T \right) = \beta T \left(\frac{\partial p}{\partial t} + \vec{v} \cdot \nabla p \right) + p\nabla \cdot \vec{v} + (\underline{\sigma} : \{\nabla \vec{v}\}) - (\nabla \cdot \vec{q}) \qquad \textbf{(4.62)}$$

Finally we relate the heat flux vector $\vec{q}$ to temperature. This can be done using Fourier's law of heat conduction [BL] which states that

$$\vec{q} = -k\nabla T \qquad \textbf{(4.63)}$$

where k is the thermal conductivity. Substituting this in Eq. (4.62) gives the final form of the energy equation:

$$\rho c_p \left(\frac{\partial T}{\partial t} + \vec{v} \cdot \nabla T \right) = \beta T \left(\frac{\partial p}{\partial t} + \vec{v} \cdot \nabla p \right) + p\nabla \cdot \vec{v} + (\underline{\sigma} : \{\nabla \vec{v}\}) + \nabla \cdot (k\nabla T) \qquad \textbf{(4.64)}$$

4.5 Boundary Conditions

The equations of continuity, momentum and energy derived in the preceding part of this Chapter are very general and describe quite general fluid flows. Our application is injection molding and it is possible to simplify the equations for this purpose. This is discussed in the following two Chapters.

To conclude this Chapter, we consider boundary conditions for the problem of injection molding. Unlike the governing equations, boundary conditions are quite specific for a particular problem. Therefore it is possible to describe boundary conditions for our problem which will hold for both the general equations in this Chapter and for the simplified equations for injection molding.

Figure (4.15) illustrates a simple mold cavity for which we discuss the required boundary conditions.

Figure 4.15: Boundary Conditions for Injection Molding Simulation

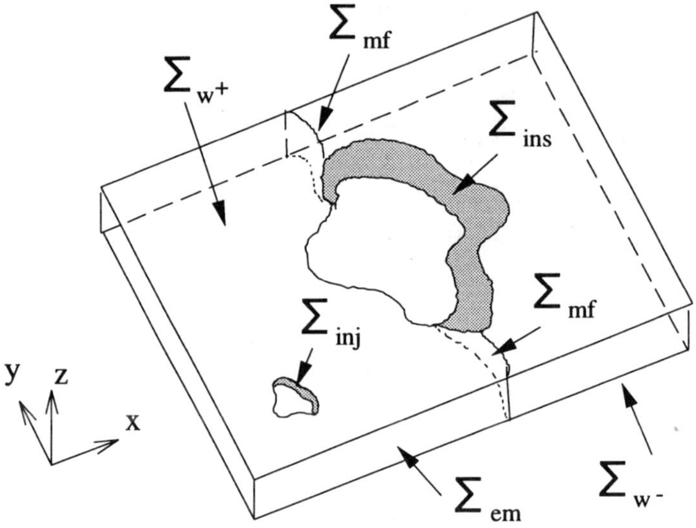

There are several surfaces on which boundary conditions need be described:

- Σ_{inj} is the surface through which the melt is injected into the cavity.
- Σ_{em} is the surface defining the edge of the mold which is in contact with the melt.
- Σ_{W+} is the top surface of the mold wall.
- Σ_{W-} is the bottom surface of the mold wall.
- Σ_{ins} is the surface defining any insert in the mold. While only one is shown in Fig. (4.15), it is possible to have any number in a real cavity.
- Σ_{mf} is the surface defining the position of the melt front.

For the injection molding problem, boundary conditions relate to the solution of the pressure and thermal distributions in the cavity. In fact, as we shall see in the next two Chapters, it is possible to combine the continuity and momentum equations into a single equation for pressure. This pressure equation is coupled to the energy equation because the material's viscosity, which affects pressure, is determined by both the temperature and shear rate. Solution of the injection molding problem therefore requires solution of the pressure and energy equations. It is convenient to discuss the pressure and thermal boundary conditions separately.

Pressure Boundary Conditions

The boundary conditions relating to pressure are

- At any impermeable boundary the pressure gradient in the normal direction to the boundary is zero. The impermeable boundaries are the mold edges, the mold walls and any mold inserts. Therefore,

$$\frac{\partial p}{\partial n} = 0 \text{ on } \Sigma_{em}, \Sigma_{w^+}, \Sigma_{w^-}, \Sigma_{ins} \tag{4.65}$$

  Physically this means that material cannot flow through the mold walls or edges.

- The melt flow rate q, or the pressure p, is specified at the surface where the melt enters the cavity. That is,

$$q = q_{inj} \text{ or } p = p_{inj} \text{ on } \Sigma_{inj} \tag{4.66}$$

  Flow analysis software usually uses a specified flow rate in the filling phase. This is obtained by dividing the volume of the part to be molded by the user specified fill time. It is possible to vary this flow rate to model injection profiling. For the packing phase it is common to specify a pressure at the point(s) of injection. This may also vary to model packing profiles.

- Assuming the pressure datum is atmospheric pressure, the pressure is zero at the melt front. That is,

$$p = 0 \text{ on } \Sigma_{mf} \tag{4.67}$$

  Since most molds are vented to allow air ahead of the melt to escape, this is physically sensible.

Temperature Boundary Conditions

The boundary conditions relating to temperature are:

- The temperature profile through the cavity thickness, $T(z)$ is prescribed for the surface through which the melt is injected. That is

$$T(z) = T_{inj}(z) \tag{4.68}$$

  Most flow analysis software assumes that the melt temperature is uniform at the point(s) of injection. In practice this is not too critical since the melt is rapidly convected into the cavity or runner system where thermal effects due to shear heating and conduction allow a temperature profile across the thickness to develop quickly.

- The temperature, T, is prescribed on all mold boundaries. A different temperature can be prescribed on each boundary. That is,

$$\begin{aligned} T &= T_{em} \text{ on } \Sigma_{em} \\ T &= T_{W+} \text{ on } \Sigma_{W+} \\ T &= T_{W-} \text{ on } \Sigma_{W-} \\ T &= T_{ins} \text{ on } \Sigma_{ins} \end{aligned} \tag{4.69}$$

  All flow analysis software allows the user to specify at least a single value of mold temperature. Often it is also possible to specify temperature on different parts of the mold. Recently it has become common to first perform a thermal analysis of the mold and its cooling system, and then use the results from this analysis to define the temperatures on the mold for flow analysis.

These boundary conditions are very general. In the next two chapters the governing equations are simplified. In some cases it is found that the simplification automatically ensures satisfaction of a particular boundary condition.

	Peter **SKU:** 1005929 **ASIN:** 1569901813 **Condition:** Used - Very Good **Order Item ID:** 169745365964361 **Condition note:** book has slight shelf wear to cover, otherwise in bright clean shape. Near Fine		**Item subtotal** $49.95 **Shipping total** $4.99 **Item total** **$54.94**

Grand total: $54.94

Returning your item:

Go to "Your Account" on Amazon.com, click "Your Orders" and then click the "seller profile" link for this order to get information about the return and refund policies that apply.

Visit https://www.amazon.com/returns to print a return shipping label. Please have your order ID ready.

Thanks for buying on Amazon Marketplace. To provide feedback for the seller please visit www.amazon.com/feedback. To contact the seller, go to Your Orders in Your Account. Click the seller's name under the appropriate product. Then, in the "Further Information" section, click "Contact the Seller."

Ship To:

AMZ Order -APTP - 1033842-1-0216
Main Address
800 Avondale Ave
1033842-1-0216
Grandview Heights, OH 43212

Buyer PO #: 1033842-1-0216

Order ID: 113-8985562-9886603

Thank you for buying from Recycle Bookstore, San Jose on Amazon Marketplace.

Shipping Address:	Order Date:	Mon, Sep 14, 2026
AMZ Order -APTP	Shipping Service:	Standard
Main Address	Buyer Name:	CollegeBooksDirect
800 Avondale Ave	Seller Name:	Recycle Bookstore, San Jose
1033842-1-0216		
Grandview Heights, OH		
43212		

Quantity	Product Details	Unit price	Order Totals

Chapter 5
Governing Equations for the Filling Phase

In the last chapter the equations governing the behaviour of a fluid motion were derived. They were:

$$\frac{\partial \rho}{\partial t} + (\nabla \cdot \rho \vec{v}) = 0 \tag{5.1}$$

$$\frac{\partial}{\partial t}(\rho \vec{v}) = \rho \vec{g} + [\nabla \cdot \underline{\sigma}] - [\nabla \cdot \rho \vec{v}\vec{v}] \tag{5.2}$$

$$\rho c_p\left(\frac{\partial T}{\partial t} + \vec{v} \cdot \nabla T\right) = \beta T\left(\frac{\partial p}{\partial t} + \vec{v} \cdot \nabla p\right) + p\nabla \cdot \vec{v} + (\underline{\sigma} : \{\nabla \vec{v}\}) - \nabla \cdot (k\nabla T) \tag{5.3}$$

These equations are quite general. They hold for all common fluids. With today's computer hardware and material testing technology it is not feasible to solve them in complicated domains like injection mold cavities. To make solution times and material data requirements more reasonable, a number of assumptions are made. In this chapter we discuss these assumptions and produce a simplified set of equations for analysis of the filling phase.

5.1 Material Assumptions

In this section we state the assumptions, used in the software, regarding the properties of the material. Each assumption allows some simplification of equations (5.1) to (5.3).

5.1.1 Melt Compressibility

Assumption: During the filling phase, the melt is assumed to be incompressible.

This assumption means that the density is constant and so, from Eq. (5.2), (see also Section3.3) we have

$$\nabla \cdot \vec{v} = 0 \tag{5.4}$$

It is important to realize that this assumption does *not* mean we ignore compressibility in the software. During the filling phase, the compressibility of the material ahead of the ram and in the feed system is accounted for. This can be justified for filling only. The assumption of incompressibility is not permissible in the packing phase. In Chapter 6 the effect of variable density on the governing equations is discussed.

Equation (5.4) allows some simplification of the momentum equation. Consider the term involving the dyadic product $\vec{v}\vec{v}$ in equation (5.2). Using the identity $(\nabla \cdot \vec{a}\vec{s}) = (\vec{a} \cdot \nabla \vec{s}) + \vec{s}(\nabla \cdot \vec{a})$ we have

$$\begin{aligned} [\nabla \cdot \rho\vec{v}\vec{v}] &= \rho[\nabla \cdot \vec{v}\vec{v}] \\ &= \rho[\vec{v} \cdot \nabla\vec{v}] + \rho\vec{v}(\nabla \cdot \vec{v}) \\ &= \rho[\vec{v} \cdot \nabla\vec{v}] \quad \text{since } (\nabla \cdot \vec{v}) = 0 \end{aligned} \tag{5.5}$$

5.1.2 Fluid Model

Assumption: The melt can be represented as a Generalized Newtonian Fluid.

This assumption implies that viscoelastic effects will be ignored. Moreover it assumes the relationship (see Section 3.4):

$$\begin{aligned} \underline{\sigma} &= -p\underline{I} + \underline{\tau} \\ &= -p\underline{I} + \eta(\dot{\gamma})\underline{\dot{\gamma}} \end{aligned} \tag{5.6}$$

Note that because of the earlier assumption that the melt is incompressible, the deviatoric strain tensor and the strain rate tensor are equal.

This assumption alters the expressions $(\nabla \cdot \underline{\sigma})$ and $(\underline{\sigma} : \nabla\vec{v})$ in equations (5.2) and (5.3) respectively.

Using equations (3.4) and (3.15), we have

$$\begin{aligned}[\nabla \cdot \underline{\sigma}] &= \left[\nabla \cdot (-p\underline{I} + \eta\underline{\dot{\gamma}})\right] \\ &= -\nabla p + \left[\nabla \cdot \eta\underline{\dot{\gamma}}\right]\end{aligned} \tag{5.7}$$

and

$$\begin{aligned}(\underline{\sigma} : \{\nabla\vec{v}\}) &= (-p\underline{I} + \underline{\tau}) : \nabla\vec{v} \\ &= (\{-p\underline{I} + \eta\underline{\dot{\gamma}}\} : \nabla\vec{v}) \\ &= (-p\underline{I} : \nabla\vec{v}) + (\eta\underline{\dot{\gamma}} : \nabla\vec{v}) \\ &= -p(\nabla \cdot \vec{v}) + (\eta\{\nabla\vec{v} + \nabla\vec{v}^{\dagger}\} : \nabla\vec{v}) \\ &= \eta(\nabla\vec{v} : \nabla\vec{v} + \nabla\vec{v}^{\dagger} : \nabla\vec{v}) \\ &\quad \text{Since fluid is incompressible} \\ &= \frac{1}{2}\eta(\nabla\vec{v} : \nabla\vec{v} + \nabla\vec{v}^{\dagger} : \nabla\vec{v} + \nabla\vec{v} : \nabla\vec{v}^{\dagger} + \nabla\vec{v}^{\dagger} : \nabla\vec{v}^{\dagger}) \\ &= \frac{1}{2}\eta(\{\nabla\vec{v} + \nabla\vec{v}^{\dagger}\} : \{\nabla\vec{v} + \nabla\vec{v}^{\dagger}\}) \\ &= \frac{1}{2}\eta(\underline{\dot{\gamma}} : \underline{\dot{\gamma}}) \\ &= \eta\dot{\gamma}^2\end{aligned} \tag{5.8}$$

where $\dot{\gamma}^2 = \frac{1}{2}\underline{\dot{\gamma}} : \underline{\dot{\gamma}}$ is the shear rate as defined by Eq. (3.22).

5.1.3 Thermal Properties of the Melt

Assumption: The thermal conductivity of the material is assumed to be constant.

Despite the fact that the thermal conductivity, k, of polymers depends on temperature (see Section 2.7.2), this assumption is enforced because of the difficulty in obtaining material data.

Using the assumption the term involving k in Eq. (5.3) becomes:

$$\nabla \cdot (k\nabla T) = k\nabla^2 T \tag{5.9}$$

5.1.4 Resulting Equations

Enforcing the assumptions above gives the following three equations:

$$(\nabla \cdot \vec{v}) = 0 \tag{5.10}$$

$$\rho \frac{\partial \vec{v}}{\partial t} = \rho \vec{g} - \nabla p + [\nabla \cdot \eta \underline{\dot{\gamma}}] - \rho[\vec{v} \cdot \nabla \vec{v}] \tag{5.11}$$

$$\rho c_p \left(\frac{\partial T}{\partial t} + \vec{v} \cdot \nabla T \right) = \beta T \left(\frac{\partial p}{\partial t} + \vec{v} \cdot \nabla p \right) + \eta \dot{\gamma}^2 + k\nabla^2 T \tag{5.12}$$

5.2 Geometric Considerations: The Cavity

Solution of equations (5.10) to (5.12) is still a formidable problem. Fortunately, further simplification is possible due to the geometry of injection molds.

To simulate the filling phase, we must deal with flow in two areas having quite different geometries:

i. the cavity

ii. the runner system.

We deal with the cavity first and then return to the runners.

5.2.1 Dimensional Analysis

In general, injection molded parts are thin walled. Flow in the cavity is therefore similar to flow in a narrow gap. Figure (5.16) shows a simple cavity geometry. For convenience a Cartesian coordinate system is used to describe the cavity. The axes are arranged so that, at any point in the cavity the x-y plane coincides with the mid-plane of the part and the z-axis points in the thickness direction.

Figure 5.16: Coordinate System for Mold Cavity

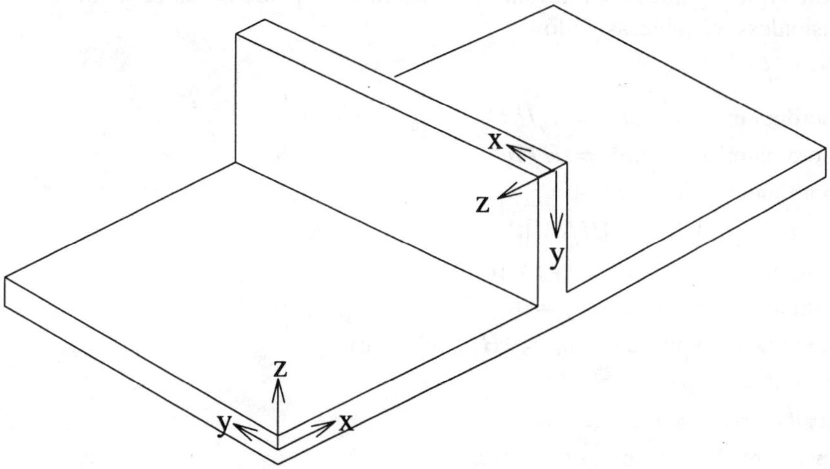

To simplify the equations, we employ a technique called dimensional analysis. Basically the idea is to obtain estimates of the order of magnitude of each term in the governing equations. Terms of sufficiently low order have little influence and so are neglected. Essential to the method is the idea of characteristic values. These are typical values of variables such as cavity length, cavity thickness, melt velocity and viscosity. Exact values are not required - order of magnitude estimates are sufficient.

Characteristic values to be used are [BW]:

Cavity thickness, $H = 10^{-3}\, m$
Cavity length, $L = H/\delta\, m$ where $\delta = H/L << 1$
Velocity of melt, $V = 10^{-1}\, m/s$
Cavity pressure, $p_0 = 10^7\, N/m^2$
Viscosity of melt, $\eta_0 = 10^4\, Ns/m^2$

Expansivity, $\beta = 10^{-3}\, 1/K$
Thermal conductivity of melt, $k = 10^{-1}\, W/mK$
Density of melt, $\rho = 10^3\, kg/m^3$
Temperature difference between mold and melt, $T_0 = 10^2\, K$

Using these typical values the relevant variables in the equations can be defined in terms of dimensionless variable as follows:

x-coordinate: $x = Lx^* = H/\delta x^*$
y-coordinate: $y = Ly^* = H/\delta y^*$
z-coordinate: $z = Hz^*$
time: $t = [L/V]t^* = [H/\delta V]t^*$
x component of velocity: $v_x = [L/T]v_x^* = Vv_x^*$
y component of velocity:$v_y = [L/T]v_y^* = Vv_y^*$
z component of velocity: $v_z = [H/T]v_z^* = \delta V v_z^*$
pressure: $p = p_0 p^*$
viscosity: $\eta = \eta_0 \eta^*$.
temperature difference: $\Delta T = T_0 \Delta T^*$.

where the starred quantities are dimensionless and of order one.

Our approach will involve substituting dimensionless variables into each equation and then estimating the order of magnitude. For example,

$$\frac{\partial v_x}{\partial x} = \left[\frac{V}{L}\right] \frac{\partial v_x^*}{\partial x^*} \tag{5.13}$$

Because the starred quantities are of order one, the order of magnitude of $\partial v_x/\partial x$ is equal to the order of magnitude of V/L. For brevity, we write:

$$\mathrm{O}\left[\frac{\partial v_x}{\partial x}\right] = \mathrm{O}\left[\frac{V}{L}\right] \tag{5.14}$$

where O[·] is read as *the order of magnitude of* [·].

Using the characteristic values given for V and L we find:

$$
\begin{aligned}
\mathrm{O}\left[\frac{\partial v_x}{\partial x}\right] &= \mathrm{O}\left[\frac{V}{L}\right] \\
&= \mathrm{O}\left[\frac{\delta V}{H}\right] \\
&= \mathrm{O}\left[\frac{10^{-1}\delta}{10^{-3}}\right] \\
&= \mathrm{O}\left[10^{2}\delta\right]
\end{aligned}
\tag{5.15}
$$

5.2.2 The Continuity Equation for the Cavity

The continuity equation (5.10) for an incompressible, generalized non-Newtonian fluid may be written in Cartesian coordinates as follows:

$$
\frac{\partial v_x}{\partial x} + \frac{\partial v_y}{\partial y} + \frac{\partial v_z}{\partial x} = 0 \tag{5.16}
$$

Using the characteristic values and non-dimensionless variables given in Section 5.2.1 we find all terms have similar order of magnitude:

$$
\begin{aligned}
O\left[\frac{\partial v_x}{\partial x}\right] &= O\left[\frac{V}{L}\right] \\
&= O\left[\frac{\delta V}{H}\right] \\
&= O\left[\frac{10^{-1}\delta}{10^{-3}}\right] \\
&= O\left[10^{2}\delta\right] \\
O\left[\frac{\partial v_y}{\partial y}\right] &= O\left[\frac{V}{L}\right] \\
&= O\left[10^{2}\delta\right] \\
O\left[\frac{\partial v_z}{\partial z}\right] &= O\left[\frac{\delta V}{H}\right] \\
&= O\left[\frac{10^{-1}\delta}{10^{-3}}\right] \\
&= O\left[10^{2}\delta\right]
\end{aligned}
\tag{5.17}
$$

Since all terms in Eq. (5.16) have similar magnitude, it is not possible to simplify the equation further.

5.2.3 The Momentum Equation for the Cavity

The momentum equation (5.11) is a vector equation. When expressed in Cartesian coordinates it yields the following three equations:

- x-component of momentum

$$\begin{aligned}\underbrace{\rho\frac{\partial v_x}{\partial t}}_{(i)} = \underbrace{\rho f_x}_{(ii)} - \underbrace{\frac{\partial p}{\partial x}}_{(iii)} + \underbrace{\frac{\partial}{\partial x}\left(2\eta\frac{\partial v_x}{\partial x}\right)}_{(iv)} + \underbrace{\frac{\partial}{\partial y}\left(\eta\left(\frac{\partial v_y}{\partial x} + \frac{\partial v_x}{\partial y}\right)\right)}_{(v)} \\ + \underbrace{\frac{\partial}{\partial z}\left(\eta\left(\frac{\partial v_x}{\partial z} + \frac{\partial v_z}{\partial x}\right)\right)}_{(vi)} - \underbrace{\rho\left(v_x\frac{\partial v_x}{\partial x} + v_y\frac{\partial v_x}{\partial y} + v_z\frac{\partial v_x}{\partial z}\right)}_{(vii)}\end{aligned} \tag{5.18}$$

- y-component of momentum

$$\begin{aligned}\underbrace{\rho\frac{\partial v_y}{\partial t}}_{(i)} = \underbrace{\rho f_y}_{(ii)} - \underbrace{\frac{\partial p}{\partial y}}_{(iii)} + \underbrace{\frac{\partial}{\partial x}\left(\eta\left(\frac{\partial v_x}{\partial y} + \frac{\partial v_y}{\partial x}\right)\right)}_{(iv)} + \underbrace{\frac{\partial}{\partial y}\left(2\eta\frac{\partial v_y}{\partial y}\right)}_{(v)} \\ + \underbrace{\frac{\partial}{\partial z}\left(\eta\left(\frac{\partial v_z}{\partial y} + \frac{\partial v_y}{\partial z}\right)\right)}_{(vi)} - \underbrace{\rho\left(v_x\frac{\partial v_y}{\partial x} + v_y\frac{\partial v_y}{\partial y} + v_z\frac{\partial v_y}{\partial z}\right)}_{(vii)}\end{aligned} \tag{5.19}$$

- z-component of momentum

$$\begin{aligned}\underbrace{\rho\frac{\partial v_z}{\partial t}}_{(i)} = \underbrace{\rho f_z}_{(ii)} - \underbrace{\frac{\partial p}{\partial z}}_{(iii)} + \underbrace{\frac{\partial}{\partial x}\left(\eta\left(\frac{\partial v_x}{\partial z} + \frac{\partial v_z}{\partial x}\right)\right)}_{(iv)} + \underbrace{\frac{\partial}{\partial y}\left(\eta\left(\frac{\partial v_y}{\partial z} + \frac{\partial v_z}{\partial y}\right)\right)}_{(v)} \\ + \underbrace{\frac{\partial}{\partial z}\left(2\eta\frac{\partial v_z}{\partial z}\right)}_{(vi)} - \underbrace{\rho\left(v_x\frac{\partial v_z}{\partial x} + v_y\frac{\partial v_z}{\partial y} + v_z\frac{\partial v_z}{\partial z}\right)}_{(vii)}\end{aligned} \tag{5.20}$$

In the above equations it is worth noting the physical significance of the terms. Term (i) is the rate of change of momentum, term (ii) represents forces due to gravity, term (iii) is the pressure gradient, terms (iv) to (vi) represent viscous forces and term (vii) inertial forces.

We will use the characteristic values and dimensionless variables of Section5.2.1 to determine the magnitudes of the terms in equations (5.18) to (5.20) and hence the relevance of each.

Below we have expanded out equations (5.18) to (5.20) and shown the order of each term. Details of the calculations may be found in Appendix C.

$$\begin{aligned}\underbrace{\left(\rho\frac{\partial v_x}{\partial t}\right)}_{(10^4\delta)} = \underbrace{(\rho f_x)}_{(10^4)} - \underbrace{\left(\frac{\partial p}{\partial x}\right)}_{(10^{10}\delta)} + \underbrace{\frac{\partial}{\partial x}\left(2\eta\frac{\partial v_x}{\partial x}\right)}_{(10^9\delta^2)} + \underbrace{\frac{\partial}{\partial y}\left(\eta\frac{\partial v_y}{\partial x}\right)}_{(10^9\delta^2)} \\ + \underbrace{\frac{\partial}{\partial y}\left(\eta\frac{\partial v_x}{\partial y}\right)}_{(10^9\delta^2)} + \underbrace{\frac{\partial}{\partial z}\left(\eta\frac{\partial v_x}{\partial z}\right)}_{(10^9)} + \underbrace{\frac{\partial}{\partial z}\left(\eta\frac{\partial v_z}{\partial x}\right)}_{(10^9\delta^2)} \\ - \underbrace{\rho v_x\frac{\partial v_x}{\partial x}}_{(10^4\delta)} - \underbrace{\rho v_y\frac{\partial v_x}{\partial y}}_{(10^4\delta)} - \underbrace{\rho v_z\frac{\partial v_x}{\partial z}}_{(10^4\delta)}\end{aligned} \tag{5.21}$$

$$\begin{aligned}\underbrace{\left(\rho\frac{\partial v_y}{\partial t}\right)}_{(10^4\delta)} = \underbrace{(\rho f_y)}_{(10^4)} - \underbrace{\left(\frac{\partial p}{\partial y}\right)}_{(10^{10}\delta)} + \underbrace{\frac{\partial}{\partial x}\left(\eta\frac{\partial v_x}{\partial y}\right)}_{(10^9\delta^2)} + \underbrace{\frac{\partial}{\partial y}\left(\eta\frac{\partial v_y}{\partial x}\right)}_{(10^9\delta^2)} \\ + \underbrace{\frac{\partial}{\partial y}\left(2\eta\frac{\partial v_y}{\partial y}\right)}_{(10^9\delta^2)} + \underbrace{\frac{\partial}{\partial z}\left(\eta\frac{\partial v_z}{\partial y}\right)}_{(10^9\delta^2)} + \underbrace{\frac{\partial}{\partial z}\left(\eta\frac{\partial v_y}{\partial z}\right)}_{(10^9)} \\ - \underbrace{\rho v_x\frac{\partial v_y}{\partial x}}_{(10^4\delta)} - \underbrace{\rho v_y\frac{\partial v_y}{\partial y}}_{(10^4\delta)} - \underbrace{\rho v_z\frac{\partial v_y}{\partial z}}_{(10^4\delta)}\end{aligned} \tag{5.22}$$

$$\underbrace{\left(\rho\frac{\partial v_z}{\partial t}\right)}_{(10^4\delta^2)} = \underbrace{(\rho f_z)}_{(10^4)} - \underbrace{\left(\frac{\partial p}{\partial z}\right)}_{(10^{10})} + \underbrace{\frac{\partial}{\partial x}\left(\eta\frac{\partial v_x}{\partial z}\right)}_{(10^9\delta)} + \underbrace{\frac{\partial}{\partial x}\left(\eta\frac{\partial v_z}{\partial x}\right)}_{(10^9\delta^3)}$$
$$+ \underbrace{\frac{\partial}{\partial y}\left(\eta\frac{\partial v_y}{\partial z}\right)}_{(10^9\delta)} + \underbrace{\frac{\partial}{\partial y}\left(\eta\frac{\partial v_z}{\partial y}\right)}_{(10^9\delta^3)} + \underbrace{\frac{\partial}{\partial z}\left(2\eta\frac{\partial v_z}{\partial z}\right)}_{(10^9\delta)} \tag{5.23}$$
$$- \underbrace{\rho v_x\frac{\partial v_z}{\partial x}}_{(10^4\delta^2)} - \underbrace{\rho v_y\frac{\partial v_z}{\partial y}}_{(10^4\delta^2)} - \underbrace{\rho v_z\frac{\partial v_z}{\partial z}}_{(10^4\delta^2)}$$

Term Number	**Description (force/unit mass)**	**Magnitude of x-cpt**	**Magnitude of y-cpt**	**Magnitude of z-cpt**
(i)	Total force component	$10^2\delta$	$10^2\delta$	$10^2\delta^2$
(ii)	Gravitational force	10^4	10^4	10^4
(iii)	Pressure gradient	$\mathbf{10^{10}}\delta$	$\mathbf{10^{10}}\delta$	$\mathbf{10^{10}}$
(iv)	Viscous force	$10^9\delta^2$	$10^9\delta^2$	$10^9\delta$
(v)	Viscous force	$10^9\delta^2$	$10^9\delta^2$	$10^9\delta^3$
(vi)	Viscous force	$10^9\delta^2$	$10^9\delta^2$	$10^9\delta$
(vii)	Viscous force	$\mathbf{10^9}$	$\mathbf{10^9}$	$10^9\delta^3$
(viii)	Viscous force	$10^9\delta^2$	$10^9\delta^2$	$10^9\delta$
(ix)	Inertial force	$10^4\delta$	$10^4\delta$	$10^4\delta^2$
(x)	Inertial force	$10^4\delta$	$10^4\delta$	$10^4\delta^2$
(xi)	Inertial force	$10^4\delta$	$10^4\delta$	$10^4\delta^2$

Since most injection molded parts are thin walled, we can expect δ to be much less than one. Typically δ will lie in the range $0.001 < \delta < 0.1$. The table above summarizes the magnitudes of the terms in the momentum equations.

Bearing in mind the likely values of δ, the most dominant terms are the pressure gradient and the viscous forces arising from the derivative in the z-direction. These terms are highlighted in the table. In particular, for $\delta = 0.1$, gravitational and inertial forces are respectively, five and six orders of magnitude *less* than the largest terms. We therefore ignore all terms less than or equal to $10^9\delta$ and so the equations for x, y and z components of momentum become, respectively:

$$\begin{aligned} \frac{\partial p}{\partial x} &= \frac{\partial}{\partial z}\left(\eta\frac{\partial v_x}{\partial z}\right) \\ \frac{\partial p}{\partial y} &= \frac{\partial}{\partial z}\left(\eta\frac{\partial v_y}{\partial z}\right) \\ \frac{\partial p}{\partial z} &= 0 \end{aligned} \tag{5.24}$$

5.2.4 The Energy Equation for the Cavity

The energy equation (5.12) may be written in Cartesian coordinates as follows:

$$\begin{aligned} \underbrace{\rho c_p\left(\frac{\partial T}{\partial t} + v_x\frac{\partial T}{\partial x} + v_y\frac{\partial T}{\partial y} + v_z\frac{\partial T}{\partial z}\right)}_{(i)} &= \underbrace{\beta T\left(\frac{\partial p}{\partial t} + v_x\frac{\partial p}{\partial x} + v_y\frac{\partial p}{\partial y} + v_z\frac{\partial p}{\partial z}\right)}_{(ii)} \\ &+ \underbrace{\eta\dot{\gamma}^2}_{(iii)} + \underbrace{k\left(\frac{\partial^2 T}{\partial x^2} + \frac{\partial^2 T}{\partial y^2} + \frac{\partial^2 T}{\partial z^2}\right)}_{(iv)} \end{aligned} \tag{5.25}$$

The physical significance of the terms is as follows. Term (i) is the total rate of energy change plus the energy due to convection in the x, y and z directions. Term (ii) is the total rate of energy change due to compression/expansion of the melt. Term (iii) is the rate of energy generation due to viscous dissipation or shear heating and term (iv) represents the rate of energy change due to conduction in the x, y and z directions.

As before, we estimate the order of magnitude of each term using the characteristic values of Section 5.2.1. Details of the calculations are in Appendix C. Expanding the equation and writing the magnitude under each term we obtain:

$$
\begin{aligned}
&\underbrace{\rho c_p \frac{\partial T}{\partial t}}_{(10^{10}\delta)} + \underbrace{\rho c_p v_x \frac{\partial T}{\partial x}}_{(10^{10}\delta)} + \underbrace{\rho c_p v_y \frac{\partial T}{\partial y}}_{(10^{10}\delta)} + \underbrace{\rho c_p v_z \frac{\partial T}{\partial z}}_{(10^{10}\delta)} \\
&\qquad = \underbrace{\beta T \frac{\partial p}{\partial t}}_{(10^{8}\delta)} + \underbrace{\beta T v_x \frac{\partial p}{\partial x}}_{(10^{8}\delta)} + \underbrace{\beta T v_y \frac{\partial p}{\partial y}}_{(10^{8}\delta)} + \underbrace{\beta T v_z \frac{\partial p}{\partial z}}_{(10^{8}\delta)} \\
&\qquad + \underbrace{\eta \dot{\gamma}^2}_{(10^{8})} + \underbrace{k \frac{\partial^2 T}{\partial x^2}}_{(10^{7}\delta^2)} + \underbrace{k \frac{\partial^2 T}{\partial y^2}}_{(10^{7}\delta^2)} + \underbrace{k \frac{\partial^2 T}{\partial z^2}}_{(10^{7})}
\end{aligned}
\tag{5.26}
$$

It can be seen that the terms involving the change of energy due to compression/expansion and conduction in the x and y directions are of significantly lower magnitude. We therefore drop these terms and obtain the following energy equation:

$$
\rho c_p \left(\frac{\partial T}{\partial t} + v_x \frac{\partial T}{\partial x} + v_y \frac{\partial T}{\partial y} + v_z \frac{\partial T}{\partial z} \right) = \eta \dot{\gamma}^2 + k \frac{\partial^2 T}{\partial z^2} \tag{5.27}
$$

where the shear rate is defined as

$$
\dot{\gamma} = \sqrt{\left(\frac{\partial v_x}{\partial z} \right)^2 + \left(\frac{\partial v_y}{\partial z} \right)^2} \tag{5.28}
$$

It is worth noting that recent work has indicated that conduction in the x and y directions can be significantly increased due to molecular orientation induced by the flow [BS]. We will neglect this effect in what follows however.

Equation (5.27) is somewhat simpler than the original equation. However, it is possible to simplify even further by ignoring the convective term in the z-direction $v_z \partial T / \partial z$. The motivation for dropping the convection term is that the velocities in the x and y directions can be obtained quite simply from the momentum equations. Neglecting the term implies that there is no convection in the z-direction. This is clearly wrong at the flow front, where so called "fountain flow" causes melt from the midstream to split and convect towards the cavity walls. That fountain flow does in fact occur is accepted. Due to cure effects, fountain flow is important for analysis of thermosetting materials. There are however conflicting views as to the importance of this for thermoplastics flow analysis. In [WD] large errors are claimed but nevertheless most researchers ignore convection in the thickness direction [HS], [HL]. Verification experiments on Moldflow software have failed to show that ignoring z-convection has a significant effect on results. For this reason we make the following assumption.

Assumption: The convective heat transfer in the z-direction may be ignored.

The energy equation therefore becomes:

$$\rho c_p \left(\frac{\partial T}{\partial t} + v_x \frac{\partial T}{\partial x} + v_y \frac{\partial T}{\partial y} \right) = \eta \dot{\gamma}^2 + k \frac{\partial^2 T}{\partial z^2} \tag{5.29}$$

5.2.5 Resulting Equations

Using assumptions about material behaviour and dimensional analysis to estimate the magnitude of the terms in each equation, we have reduced equations (5.10) to (5.12) to the following:

Continuity Equation

$$\frac{\partial v_x}{\partial x} + \frac{\partial v_y}{\partial y} + \frac{\partial v_z}{\partial z} = 0 \tag{5.30}$$

Momentum Equations

$$\begin{aligned} \frac{\partial p}{\partial x} &= \frac{\partial}{\partial z}\left(\eta \frac{\partial v_x}{\partial z} \right) \\ \frac{\partial p}{\partial y} &= \frac{\partial}{\partial z}\left(\eta \frac{\partial v_y}{\partial z} \right) \\ \frac{\partial p}{\partial z} &= 0 \end{aligned} \tag{5.31}$$

Energy Equation

$$\rho c_p \left(\frac{\partial T}{\partial t} + v_x \frac{\partial T}{\partial x} + v_y \frac{\partial T}{\partial y} \right) = \eta \dot{\gamma}^2 + k \frac{\partial^2 T}{\partial z^2} \tag{5.32}$$

5.3 Simplification by Mathematical Analysis

Thus far we have used both material properties and the cavity geometry to simplify the governing equations for the filling phase. Further simplification is available by integrating the momentum and continuity equations.

5.3.1 Integration of the Momentum Equations

From the last of the momentum equations (5.31) we see that the pressure is a function of x and y only. For this reason it is convenient to integrate the momentum equations across the thickness with the aim of obtaining expressions for the velocities v_x and v_y.

Before doing this however we note that, in injection molding, the material is in both solid and liquid phases. The melt is "sandwiched" between a layer of frozen material which adheres to the cavity walls. Our governing equations were derived for fluids and it is inappropriate to apply these to the frozen layer. We therefore use the concept of *no-flow temperature* to distinguish between solid and liquid phases. We define the no-flow temperature of the material to be that temperature at which the material may be considered solid. We summaries the above in the following assumption:

> **Assumption:** During filling there exists a region called the *frozen layer* in which the velocities v_x, v_y and v_z are zero. Moreover, the frozen layer is defined to be the material which lies below the no-flow temperature.

That such a frozen layer does exist has been well documented in the literature. What is contentious is how one defines it. While Moldflow uses the concept of a *no-flow temperature*, other flow analysis packages may extrapolate viscosity data (see Section 2.7.3).

Figure (5.17) defines some notation to be used. Referring to the figure, the centre-line is at z=0 and the location of the interface between the frozen layer and the melt at a point (x, y) is at $z = h^+(x, y)$ and $z = h^-(x, y)$. Cavity thickness is $2H$ and the thickness of the melt region at any point (x, y) is $h^+ - h^-$. Note that no assumption regarding symmetry of the frozen layer thickness has yet been made. Therefore it is possible that $|h^+| \neq |h^-|$.

Figure 5.17: Frozen Layer Thickness

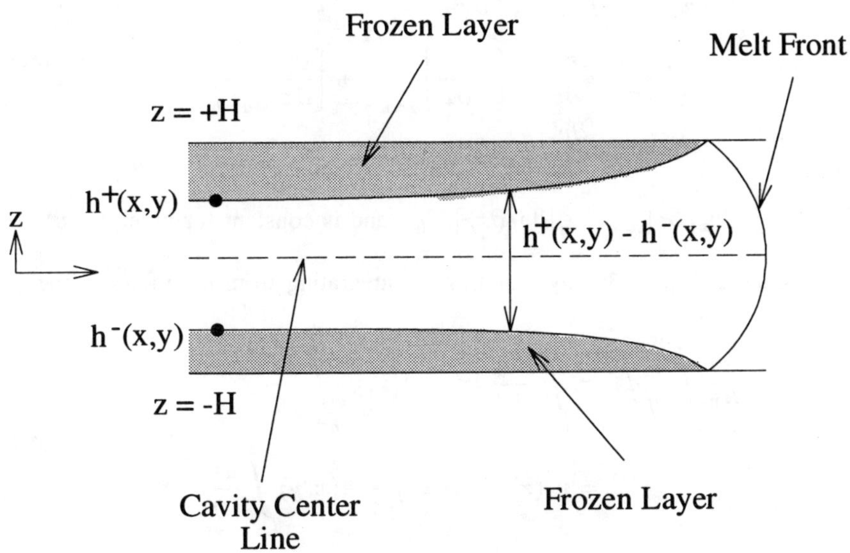

For convenience we rewrite the momentum equations below:

$$\frac{\partial p}{\partial x} = \frac{\partial}{\partial z}\left(\eta \frac{\partial v_x}{\partial z}\right) \tag{5.33}$$

$$\frac{\partial p}{\partial y} = \frac{\partial}{\partial z}\left(\eta \frac{\partial v_y}{\partial z}\right) \tag{5.34}$$

$$\frac{\partial p}{\partial z} = 0 \tag{5.35}$$

To begin, we integrate Eq. (5.33) with respect to z from h^- to z and obtain:

$$\begin{aligned}\int_{h^-}^{z} \frac{\partial p}{\partial x} dz' &= \int_{h^-}^{z} \frac{\partial}{\partial z'} \left(\eta \frac{\partial v_x}{\partial z'} \right) dz' \\ \frac{\partial p}{\partial x} z &= \eta \frac{\partial v_x}{\partial z} - \left[\eta \frac{\partial v_x}{\partial z} \right]_{z=h^-} + \left[\frac{\partial p}{\partial x} z \right]_{z=h^-} \\ &= \eta \frac{\partial v_x}{\partial z} - A(x,y)\end{aligned} \tag{5.36}$$

where $A(x,y) = [\eta \partial v_x / \partial z]_{z=h^-} - [\partial p / \partial x z]_{z=h^-}$ and is constant for given (x,y).

Dividing both sides of Eq. (5.36) by η and again integrating from h^- to z we get:

$$\begin{aligned}\frac{\partial p}{\partial x} \int_{h^-}^{z} \frac{z'}{\eta} dz' &= \int_{h^-}^{z} \frac{\partial v_x}{\partial z'} dz' - A(x,y) \int_{h^-}^{z} \frac{dz'}{\eta} \\ &= v_x(z) - v_x(h^-) - A(x,y) \int_{h^-}^{z} \frac{dz'}{\eta}\end{aligned} \tag{5.37}$$

At the interface formed by the frozen layer and the melt, where $z = h^-$ or $z = h^+$, we have $v_x = 0$. This is called a *no-slip* condition. Applying the no-slip condition and rearranging Eq. (5.37) we obtain:

$$v_x(z) = \frac{\partial p}{\partial x} \int_{h^-}^{z} \frac{z'}{\eta} dz' + A(x,y) \int_{h^-}^{z} \frac{dz'}{\eta} \tag{5.38}$$

Carrying out the above operations on Eq. (5.34) we obtain a similar expression for the y-component of velocity:

$$v_y(z) = \frac{\partial p}{\partial y} \int_{h^-}^{z} \frac{z'}{\eta} dz' + B(x,y) \int_{h^-}^{z} \frac{dz'}{\eta} \tag{5.39}$$

where $B(x,y) = [\eta \partial v_y / \partial z]_{z=h^-} - [\partial p / \partial y z]_{z=h^-}$ and is constant for given (x,y).

Applying the no-slip condition at $z = h^+$ to Eq. (5.38) we obtain:

$$0 = \frac{\partial p}{\partial x} \int_{h^-}^{h^+} \frac{z'}{\eta} dz' + A(x,y) \int_{h^-}^{h^+} \frac{dz'}{\eta} \qquad \textbf{(5.40)}$$

and so

$$\begin{aligned} A(x,y) &= -\frac{\partial p}{\partial x} \left\{ \frac{\int_{h^-}^{h^+} \frac{z'}{\eta} dz'}{\int_{h^-}^{h^+} \frac{dz'}{\eta}} \right\} \\ &= -\frac{\partial p}{\partial x} C(x,y) \end{aligned} \qquad \textbf{(5.41)}$$

where we have defined the constant

$$C(x,y) = \frac{\int_{h^-}^{h^+} \frac{z'}{\eta} dz'}{\int_{h^-}^{h^+} \frac{dz'}{\eta}} \qquad \textbf{(5.42)}$$

Similarly, using Eq. (5.39) we obtain

$$\begin{aligned} B(x,y) &= -\frac{\partial p}{\partial y} \left\{ \frac{\int_{h^-}^{h^+} \frac{z'}{\eta} dz'}{\int_{h^-}^{h^+} \frac{dz'}{\eta}} \right\} \\ &= -\frac{\partial p}{\partial y} C(x,y) \end{aligned} \qquad \textbf{(5.43)}$$

Using these results, equations (5.38) and (5.39) may be written, respectively, as:

$$v_x(z) = \frac{\partial p}{\partial x} \left[\int_{h^-}^{z} \frac{z'}{\eta} dz' - C(x,y) \int_{h^-}^{z} \frac{dz'}{\eta} \right] \qquad \textbf{(5.44)}$$

$$v_y(z) = \frac{\partial p}{\partial y}\left[\int_{h^-}^{z} \frac{z'}{\eta}dz' - C(x,y)\int_{h^-}^{z} \frac{dz'}{\eta}\right] \tag{5.45}$$

We now define average velocities $\overline{v}_x$ and $\overline{v}_y$ as follows:

$$\overline{v}_x = \frac{1}{h^+ - h^-}\int_{h^-}^{h^+} v_x(z)dz \tag{5.46}$$

$$\overline{v}_y = \frac{1}{h^+ - h^-}\int_{h^-}^{h^+} v_y(z)dz \tag{5.47}$$

Substituting Eq. (5.44) into Eq. (5.46) gives:

$$\overline{v}_x = \frac{1}{h^+ - h^-}\frac{\partial p}{\partial x}\left[\int_{h^-}^{h^+}\int_{h^-}^{z} \frac{z'}{\eta}\,dz'dz - C(x,y)\int_{h^-}^{h^+}\int_{h^-}^{z} \frac{dz'}{\eta}\,dz\right] \tag{5.48}$$

This integral may be evaluated using integration by parts. To evaluate

$$\int_{h^-}^{h^+}\int_{h^-}^{z} \frac{z'}{\eta}dz'dz \tag{5.49}$$

we set $u = z$ and $v = \int_{h^-}^{z} \frac{z'}{\eta}dz'$, then

$$\begin{aligned}
\int_{h^-}^{h^+}\int_{h^-}^{z}\frac{z'}{\eta}dz'dz &= \int_{h^-}^{h^+} v\frac{du}{dz}dz \\
&= [vu]_{z=h^-}^{z=h^+} - \int_{h^-}^{h^+} u\frac{dv}{dz}dz \\
&= \left[z\int_{h^-}^{z}\frac{z'}{\eta}dz'\right]_{z=h^-}^{z=h^+} - \int_{h^-}^{h^+}\frac{z'^2}{\eta}dz' \\
&= h^+\int_{h^-}^{h^+}\frac{z'}{\eta}dz' - \int_{h^-}^{h^+}\frac{z'^2}{\eta}dz'
\end{aligned} \tag{5.50}$$

Similarly, to evaluate

$$\int_{h^-}^{h^+}\int_{h^-}^{z}\frac{dz'}{\eta}dz \tag{5.51}$$

we set $u = z$ and $v = \int_{h^-}^{z}\frac{dz'}{\eta}$, then

$$\begin{aligned}
\int_{h^-}^{h^+}\int_{h^-}^{z}\frac{dz'}{\eta}dz &= \int_{h^-}^{h^+} v\frac{du}{dz}dz \\
&= [vu]_{z=h^-}^{z=h+} - \int_{h^-}^{h^+} u\frac{dv}{dz}dz \\
&= \left[z\int_{h^-}^{z}\frac{dz'}{\eta}\right]_{z=h^-}^{z=h^+} - \int_{h^-}^{h^+}\frac{z'}{\eta}dz' \\
&= h^+\int_{h^-}^{h^+}\frac{dz'}{\eta} - \int_{h^-}^{h^+}\frac{z'}{\eta}dz'
\end{aligned} \tag{5.52}$$

Substituting equations (5.50) and (5.52) into Eq. (5.48) and using Eq. (5.42) gives

$$\begin{aligned}
\overline{v}_x &= \frac{1}{h^+ - h}\frac{\partial p}{\partial x}\left[h^+ \int_{h^-}^{h^+} \frac{z'}{\eta} dz' - \int_{h^-}^{h^+} \frac{z'^2}{\eta} dz - C(x,y)\left\{ h^+ \int_{h^-}^{h^+} \frac{dz'}{\eta} - \int_{h^-}^{h^+} \frac{z'}{\eta} dz' \right\}\right] \\
&= \frac{1}{h^+ - h}\frac{\partial p}{\partial x}\left[h^+ \int_{h^-}^{h^+} \frac{z'}{\eta} dz' - \int_{h^-}^{h^+} \frac{z'^2}{\eta} dz - h^+ \int_{h^-}^{h^+} \frac{z'}{\eta} dz' + \frac{\left(\int_{h^-}^{h^+} \frac{z'}{\eta} dz' \right)^2}{\int_{h^-}^{h^+} \frac{dz'}{\eta}} \right] \\
&= \frac{1}{h^+ - h^-}\frac{\partial p}{\partial x}\left[- \int_{h^-}^{h^+} \frac{z'^2}{\eta} dz' + \frac{\left(\int_{h^-}^{h^+} \frac{z'}{\eta} dz' \right)^2}{\int_{h^-}^{h^+} \frac{dz'}{\eta}} \right] \\
&= \frac{-2S_2}{h^+ - h^-}\frac{\partial p}{\partial x}
\end{aligned} \tag{5.53}$$

where we have defined

$$S_2 = \frac{1}{2}\left\{ \int_{h^-}^{h^+} \frac{z'^2}{\eta} dz' - \frac{\left(\int_{h^-}^{h^+} \frac{z'}{\eta} dz' \right)^2}{\int_{h^-}^{h^+} \frac{dz'}{\eta}} \right\} \tag{5.54}$$

The quantity S_2 is called the fluidity. We use the subscript to indicate that it is associated with flow in a channel which is essentially 2–dimensional.

The average y-component of velocity may be found similarly by substituting Eq. (5.45) in Eq. (5.47). In this way we find,

$$\overline{v}_y = \frac{-2S_2}{h^+ - h^-}\frac{\partial p}{\partial y} \tag{5.55}$$

5.3.2 Integration of the Continuity Equation

In the filling phase, the continuity equation is

$$\frac{\partial v_x}{\partial x} + \frac{\partial v_y}{\partial y} + \frac{\partial v_z}{\partial z} = 0 \qquad \textbf{(5.56)}$$

Integration of the left hand side of this equation, with respect to z, between $z = -H$ and $z = +H$ gives,

$$\int_{H^-}^{H^+} \frac{\partial v_x}{\partial x} dz + \int_{H^-}^{H^+} \frac{\partial v_y}{\partial y} dz + \int_{H^-}^{H^+} \frac{\partial v_z}{\partial z} dz = \int_{h^-}^{h^+} \frac{\partial v_x}{\partial x} dz + \int_{h^-}^{h^+} \frac{\partial v_y}{\partial y} dz + [v_z]_{z=-H}^{z=+H}$$

(since $v_x = v_y = v_z = 0$ in the frozen layer.)

$$= \frac{\partial}{\partial x} \int_{h^-}^{h^+} v_x dz + \frac{\partial}{\partial y} \int_{h^-}^{h^+} v_y dz$$

$$= \frac{\partial}{\partial x}((h^+ - h^-)\overline{v}_x) + \frac{\partial}{\partial y}((h^+ - h^-)\overline{v}_y)$$

$$= -\frac{\partial}{\partial x}\left(2S_2 \frac{\partial p}{\partial x}\right) - \frac{\partial}{\partial y}\left(2S_2 \frac{\partial p}{\partial y}\right) \qquad \textbf{(5.57)}$$

where, in the last two expressions, we have used the definition and values of average velocities given in equations (5.46), (5.47), (5.53) and (5.55).

Substituting the above in Eq. (5.56) the we obtain the following equation:

$$\frac{\partial}{\partial x}\left(S_2 \frac{\partial p}{\partial x}\right) + \frac{\partial}{\partial y}\left(S_2 \frac{\partial p}{\partial y}\right) = 0 \qquad \textbf{(5.58)}$$

Equation (5.58) is a single equation for pressure that combines the momentum and continuity equations.

5.4 Simplification Due to Symmetry

Equation (5.58) can be simplified further if we assume that the flow field is symmetric about the cavity centre line.

In addition to the simplifying Eq. (5.58), symmetry of the flow field also means that calculations need only be performed for one half of the cavity thickness, enabling both considerable gain in computational speed and reduction of storage requirements. We therefore accept this assumption.

> **Assumption:** The flow field is assumed to be symmetrical about the cavity centre line.

This assumption implies that the frozen layer is also symmetrical about the cavity centre line. In some situations, this is not valid. For example near a direct sprue, at sharp corners and in molds in which the temperature between mold halves is substantial.

Asymmetric flow near the sprue and at corners is fortunately a local effect and so little error should be introduced by the assumption.

Temperature variation between mold halves is frequently encountered in real molds. Due to the low thermal conductivity of plastics, the temperature variation must be quite high to induce significant asymmetry and so again little error should be produced.

As a result of this assumption we have the following:

1. The frozen layer thickness is symmetrical about the centre line, that is

$$\begin{aligned} |h^+| &= |h^-| \\ &= h \end{aligned} \quad \textbf{(5.59)}$$

2. The velocity profile across the cavity is symmetrical about the cavity centre line ($z = 0$), therefore

$$\frac{\partial v_x}{\partial z} = \frac{\partial v_y}{\partial z} = 0 \quad \text{at } z = 0 \quad \textbf{(5.60)}$$

3. The temperature field is symmetrical about the cavity centre line ($z = 0$), therefore

$$\frac{\partial T}{\partial z} = 0 \quad \text{at } z = 0 \quad \textbf{(5.61)}$$

4. The viscosity field is symmetrical about the cavity centre line ($z = 0$). This means that

$$\begin{aligned}\int_{h^-}^{h^+} \frac{z'}{\eta} dz' &= \int_{-h}^{h} \frac{z'}{\eta} dz' \\ &= \int_{0}^{h} \frac{z'}{\eta} dz' + \int_{-h}^{0} \frac{z'}{\eta} dz' \\ &= \int_{0}^{h} \frac{z'}{\eta} dz' - \int_{0}^{h} \frac{z'}{\eta} dz' \\ &= 0\end{aligned} \tag{5.62}$$

and so, from Eq. (5.54) we have,

$$\begin{aligned}S_2 &= \frac{1}{2} \int_{h^-}^{h^+} \frac{z'^2}{\eta} dz' \\ &= \frac{1}{2} \int_{-h}^{h} \frac{z'^2}{\eta} dz' \\ &= \int_{0}^{h} \frac{z'^2}{\eta} dz'\end{aligned} \tag{5.63}$$

5.5 Equations for Runners

The assumptions regarding material behaviour made in Section 5.1 are also valid for the runner system. Our starting point in this section is therefore the equations of continuity, momentum and energy given in Section 5.1.4.

Just as for the case of the cavity, geometric considerations allow further simplification of the governing equations. In what follows the governing equations for circular runners in the filling phase are simplified.

Figure (5.18) shows the geometry of the circular runner and its associated coordinate system. Cylindrical coordinates (r, θ, x) are used since they are natural for such geometries.

Figure 5.18: Geometry and Coordinate System for Circular Runners

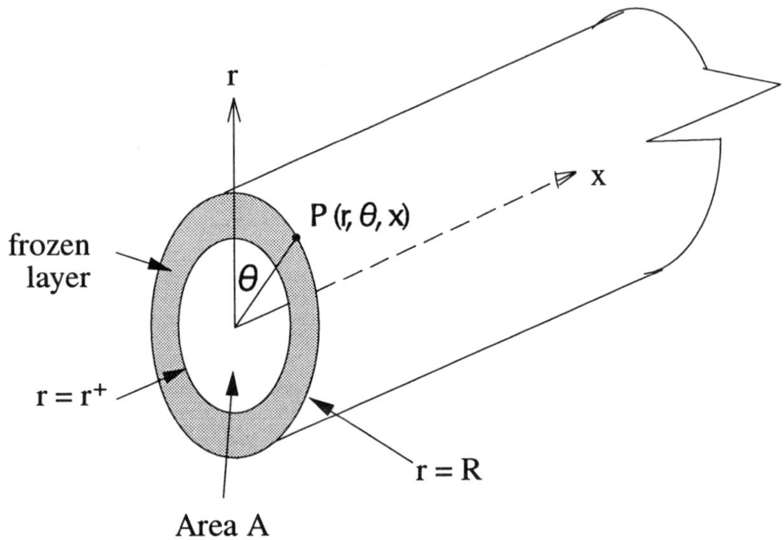

To simplify the analysis we make the following assumptions:

> **Assumption:** The temperature and flow fields in runners with circular cross-section are assumed symmetrical about the longitudinal axis of the runner.

This assumption means that no quantity will depend on θ. In particular, it means that the interface between the melt and frozen layer can be represented as a function of r and x only. In Figure (5.18) the interface is at the position $r = r^+$.

5.5.1 The Continuity Equation for Runners

In cylindrical coordinates, velocity $\vec{v}(r, \theta, x)$ has components v_r, v_θ and v_x.

Using the relationships in Appendix B, Eq. (5.10) becomes:

$$\begin{aligned} 0 &= \nabla \cdot \vec{v} \\ &= \frac{1}{r}\frac{\partial}{\partial r}(r v_r) + \frac{1}{r}\frac{\partial v_\theta}{\partial \theta} + \frac{\partial v_x}{\partial x} \end{aligned} \tag{5.64}$$

Due to our assumption regarding symmetry about the axis of the runner, this simplifies to

$$0 = \frac{1}{r}\frac{\partial}{\partial r}(r v_r) + \frac{\partial v_x}{\partial x} \tag{5.65}$$

5.5.2 The Momentum Equations for Runners

In cylindrical coordinates the momentum equation (5.11) gives the following three scalar equations:

- r-component of momentum

$$\begin{aligned} \rho\frac{\partial v_r}{\partial t} &= \rho f_r - \frac{\partial p}{\partial r} \\ &\quad + \frac{2}{r}\frac{\partial}{\partial r}\left(r\eta\frac{\partial v_r}{\partial r}\right) + \frac{\partial}{\partial x}\left(\eta\left(\frac{\partial v_r}{\partial x} + \frac{\partial v_x}{\partial r}\right)\right) - \frac{2v_r}{r^2} \\ &\quad - \rho\left(v_r\frac{\partial v_r}{\partial r} - \frac{v_\theta^2}{r} + v_x\frac{\partial v_r}{\partial x}\right) \end{aligned} \tag{5.66}$$

- θ-component of momentum

$$\begin{aligned} \rho\frac{\partial v_\theta}{\partial t} &= \rho f_\theta - \frac{1}{r}\frac{\partial p}{\partial \theta} \\ &\quad + \frac{1}{r^2}\frac{\partial}{\partial r}\left(r^3\eta\frac{\partial}{\partial r}\left(\frac{v_\theta}{r}\right)\right) + \frac{\partial}{\partial x}\left(\eta\frac{\partial v_\theta}{\partial x}\right) \\ &\quad - \rho\left(v_r\frac{\partial v_\theta}{\partial r} - \frac{v_\theta v_r}{r} + v_x\frac{\partial v_\theta}{\partial x}\right) \end{aligned} \tag{5.67}$$

- x-component of momentum

$$\begin{aligned} \rho\frac{\partial v_x}{\partial t} &= \rho f_x - \frac{\partial p}{\partial x} \\ &\quad + \frac{1}{r}\frac{\partial}{\partial r}\left(r\eta\left(\frac{\partial v_r}{\partial x} + \frac{\partial v_x}{\partial r}\right)\right) + 2\frac{\partial}{\partial x}\left(\eta\frac{\partial v_x}{\partial x}\right) \\ &\quad - \rho\left(v_r\frac{\partial v_x}{\partial r} + v_x\frac{\partial v_x}{\partial x}\right) \end{aligned} \tag{5.68}$$

To obtain the above equations the cylindrical coordinate forms of the vector operators given in Appendix B were substituted in Eq. (5.11). Due to the symmetry assumption, all derivatives with respect to θ have been set to zero.

The relative magnitudes of each of the terms in equations (5.66) to (5.68) may be established using dimensional analysis as was done in Section 5.2.3. The equations then simplify to

$$\begin{aligned} \frac{\partial p}{\partial r} &= 0 \\ \frac{\partial p}{\partial \theta} &= 0 \\ \frac{\partial p}{\partial x} &= \frac{1}{r}\frac{\partial}{\partial r}\left(r\eta\frac{\partial v_x}{\partial r}\right) \end{aligned} \tag{5.69}$$

5.5.3 The Energy Equation for Runners

Using the gradient and Laplacian operators in cylindrical coordinates given in Appendix B and assuming symmetry about the runner axis, the energy equation (5.12) becomes:

$$\begin{aligned} \rho c_p\left(\frac{\partial T}{\partial t} + v_r\frac{\partial T}{\partial r} + v_x\frac{\partial T}{\partial x}\right) &= \beta T\left(\frac{\partial p}{\partial t} + v_r\frac{\partial p}{\partial r} + v_x\frac{\partial p}{\partial x}\right) \\ &+ \eta\dot{\gamma}^2 + k\left(\frac{1}{r}\frac{\partial}{\partial r}\left(r\frac{\partial T}{\partial r}\right) + \frac{\partial^2 T}{\partial x^2}\right) \end{aligned} \tag{5.70}$$

Once again, dimensional analysis may be used to determine the relative magnitude of the terms in this equation. It turns out that the pressure terms and the conduction term in the x direction may be ignored. Therefore Eq. (5.70) simplifies to

$$\rho c_p\left(\frac{\partial T}{\partial t} + v_r\frac{\partial T}{\partial r} + v_x\frac{\partial T}{\partial x}\right) = \eta\dot{\gamma}^2 + \frac{k}{r}\frac{\partial}{\partial r}\left(r\frac{\partial T}{\partial r}\right) \tag{5.71}$$

where

$$\dot{\gamma} = \frac{\partial v_x}{\partial r} \tag{5.72}$$

Further simplification is possible if we ignore the convective heat transfer in the radial direction. Such an assumption is in error at the flow front where fountain flow is occurring. However, this is a transient effect. Once the flow front has passed a point, the flow tends to be laminar and the assumption is valid.

Assumption: The convected heat transfer in the radial direction may be ignored.

Subject to this assumption we obtain the final form of the energy equation for runners:

$$\rho c_p\left(\frac{\partial T}{\partial t} + v_x\frac{\partial T}{\partial x}\right) = \eta\dot{\gamma}^2 + \frac{k}{r}\frac{\partial}{\partial r}\left(r\frac{\partial T}{\partial r}\right) \tag{5.73}$$

5.5.4 Resulting Equations for Runners

After making assumptions regarding the material and using the effects of geometry, the equations governing the flow of a polymer melt in a runner of circular cross-section are:

- Continuity:

$$0 = \frac{1}{r}\frac{\partial}{\partial r}(rv_r) + \frac{\partial v_x}{\partial x} \tag{5.74}$$

- Momentum:

$$\frac{\partial p}{\partial x} = \frac{1}{r}\frac{\partial}{\partial r}\left(r\eta\frac{\partial v_x}{\partial r}\right) \tag{5.75}$$

- Energy:

$$\rho c_p\left(\frac{\partial T}{\partial t} + v_x\frac{\partial T}{\partial x}\right) = \eta\dot{\gamma}^2 + \frac{k}{r}\frac{\partial}{\partial r}\left(r\frac{\partial T}{\partial r}\right) \tag{5.76}$$

5.5.5 Integration of the Momentum Equation for Runners

Equation (5.75) may be integrated twice with respect to r to obtain an expression for the x-component of velocity, v_x.

Multiplying both sides of Eq. (5.75) by r and integrating with respect to r from $r = 0$ to r gives

$$\frac{\partial p}{\partial x}\int_0^r r'dr' = \int_0^r \frac{\partial}{\partial r'}\left(r'\eta\frac{\partial v_x}{\partial r'}\right)dr'$$
$$\frac{1}{2}\frac{\partial p}{\partial x}r^2 = r\eta\frac{\partial v_x}{\partial r} \tag{5.77}$$

Dividing both sides by $r\eta$ and integrating again from 0 to r we obtain:

$$\frac{1}{2}\frac{\partial p}{\partial x}\int_0^r \frac{r'}{\eta}dr' = \int_0^r \frac{\partial v_x}{\partial r'}dr'$$
$$= v_x(r) - v_x(0) \tag{5.78}$$

Let r^+ denote the value of r at the interface of the frozen layer and melt. Since v_x is zero in the frozen layer, we have from Eq. (5.78),

$$v_x(0) = v_x(r^+) - \frac{1}{2}\frac{\partial p}{\partial x}\int_0^{r^+}\frac{r'}{\eta}dr'$$
$$= -\frac{1}{2}\frac{\partial p}{\partial x}\int_0^{r^+}\frac{r'}{\eta}dr' \tag{5.79}$$

Hence Eq. (5.78) may be written as

$$v_x(r) = \frac{1}{2}\frac{\partial p}{\partial x}\left\{\int_0^{r}\frac{r'}{\eta}dr' - \int_0^{r^+}\frac{r'}{\eta}dr'\right\} \tag{5.80}$$

An average velocity in the x-direction, $\overline{v}_x(x)$ may be defined as the total flow rate through the runner cross-section divided by the area, A_c, of the melt channel. That is,

$$\overline{v}_x(x) = \frac{1}{A_c}\int_{A_c} v_x(r)dA \tag{5.81}$$

From Figure (5.18) and our symmetry assumption we have, $A = \pi r^2$ and so $dA/dr = 2\pi r$. Hence,

$$\begin{aligned}\overline{v}_x &= \frac{1}{A_c}\int_{A_c} v_x(r)dA \\ &= \frac{1}{\pi r^{+2}}\int_0^{r^+} 2\pi v_x(r) r dr \\ &= \frac{2}{r^{+2}}\int_0^{r^+} v_x(r) r dr\end{aligned} \tag{5.82}$$

Substituting Eq. (5.80) in the above equation gives:

$$\begin{aligned}\overline{v}_x &= \frac{1}{A_c}\int_{A_c} v_x(r)dA \\ &= \frac{1}{\pi r^{+2}}\int_0^{r^+} 2\pi v_x(r) r dr \\ &= \frac{2}{r^{+2}}\int_0^{r^+} v_x(r) r dr \\ &= \frac{1}{r^{+2}}\frac{\partial p}{\partial x}\left\{\int_0^{r^+}\left(\int_0^{r}\frac{r'}{\eta}dr'\right) r dr - \int_0^{r^+}\left(\int_0^{r^+}\frac{r'}{\eta}dr'\right) r dr\right\} \\ &= \frac{1}{r^{+2}}\frac{\partial p}{\partial x}\left\{\int_0^{r^+}\left(\int_0^{r}\frac{r'}{\eta}dr'\right) r dr - \left(\int_0^{r^+} r dr\right)\int_0^{r^+}\frac{r'}{\eta}dr'\right\} \\ &= \frac{1}{r^{+2}}\frac{\partial p}{\partial x}\left\{\int_0^{r^+}\left(\int_0^{r}\frac{r'}{\eta}dr'\right) r dr - \frac{1}{2}r^{+2}\int_0^{r^+}\frac{r'}{\eta}dr'\right\}\end{aligned} \tag{5.83}$$

The double integral on the RHS of (5.83) may be evaluated using integration by parts. Set

$$
\begin{aligned}
u &= \frac{1}{2}r^2 \\
v &= \int_0^r \frac{r'}{\eta} dr'
\end{aligned}
\tag{5.84}
$$

then

$$
\begin{aligned}
\int_0^{r^+} \left(\int_0^r \frac{r'}{\eta} dr' \right) r dr' &= \int_0^{r^+} v \frac{\partial u}{\partial r} dr \\
&= vu\Big|_{r=0}^{r=r^+} - \int_0^{r^+} u \frac{\partial v}{\partial r} dr \\
&= \frac{1}{2} r^{+^2} \int_0^{r^+} \frac{r'}{\eta} dr' - \frac{1}{2} \int_0^{r^+} \frac{r'^3}{\eta} dr'
\end{aligned}
\tag{5.85}
$$

Substituting this result in Eq. (5.83) gives:

$$
\begin{aligned}
\overline{v}_x(x) &= -\frac{1}{2r^{+^2}} \frac{\partial p}{\partial x} \left\{ \int_0^{r^+} \frac{r'^3}{\eta} dr' \right\} \\
&= -\frac{S_1}{r^+} \frac{\partial p}{\partial x}
\end{aligned}
\tag{5.86}
$$

where the fluidity, S_1, is defined by

$$
S_1 = \frac{1}{2r^+} \int_0^{r^+} \frac{r'^3}{\eta} dr' \tag{5.87}
$$

Note that the subscript indicates that the fluidity is for one dimensional flow.

5.5.6 Integration of the Continuity Equation for Runners

The continuity equation is from Eq. (5.65)

$$\frac{1}{r}\frac{\partial}{\partial r}(rv_r) + \frac{\partial v_x}{\partial x} = 0 \tag{5.88}$$

Integrating the left hand side of this equation over the area of the melt channel we obtain

$$\begin{aligned}
\int_{A_c} \left\{ \frac{1}{r}\frac{\partial}{\partial r}(rv_r) + \frac{\partial v_x}{\partial x} \right\} dA &= 2\pi \int_0^{r^+} \left\{ \frac{1}{r}\frac{\partial}{\partial r}(rv_r) + \frac{\partial v_x}{\partial x} \right\} r dr \\
&= 2\pi \left\{ \int_0^{r^+} \frac{\partial}{\partial r}(rv_r) dr + \int_0^{r^+} \frac{\partial v_x}{\partial x} r dr \right\} \\
&= 2\pi \left\{ (rv_r)|_{r=0}^{r=r^+} + \frac{\partial}{\partial x} \int_0^{r^+} v_x(r) r dr \right\} \\
&= 2\pi \frac{\partial}{\partial x} \left(\frac{\overline{v}_x(x) {r^+}^2}{2} \right) \\
&= -\pi \frac{\partial}{\partial x} \left(r^+ S_1 \frac{\partial p}{\partial x} \right)
\end{aligned} \tag{5.89}$$

where, in the last two steps we have used the definition of the average velocity given in Eq. (5.82) and Eq. (5.86) respectively. The continuity equation therefore reduces to

$$\frac{\partial}{\partial x}\left(r^+ S_1 \frac{\partial p}{\partial x} \right) = 0 \tag{5.90}$$

Our original equations for continuity (equation (5.65)) and momentum (equation (5.69)) have been reduced to a single equation for pressure. The equation is still quite complicated since the fluidity depends on the viscosity, which in turn is affected by pressure.

5.6 Summary

Starting with the very general equations derived in Chapter 4, a number of assumptions were made to simplify the equations for fluid flow in the filling phase. These assumptions related to both the material and the geometry of the region in which the equations were applied.

The resulting equations are:

- **Pressure Equation for the Cavity**

$$\frac{\partial}{\partial x}\left(S_2\frac{\partial p}{\partial x}\right)+\frac{\partial}{\partial y}\left(S_2\frac{\partial p}{\partial y}\right)=0 \tag{5.91}$$

where $S_2 = \int_0^h \frac{z'^2}{\eta}dz'$

- **Energy Equation for Cavity**

$$\rho c_p\left(\frac{\partial T}{\partial t}+v_x\frac{\partial T}{\partial x}+v_y\frac{\partial T}{\partial x}\right)=\eta\dot{\gamma}^2+k\frac{\partial^2 T}{\partial z^2} \tag{5.92}$$

- **Pressure Equation for Circular Runners**

$$\frac{\partial}{\partial x}\left(r^+S_1\frac{\partial p}{\partial x}\right)=0 \tag{5.93}$$

where $S_1 = \frac{1}{2r^+}\int\limits_0^{r^+}\frac{r'^3}{\eta}\,dr'$

- **Energy Equation for Circular Runners**

$$\rho c_p\left(\frac{\partial T}{\partial t}+v_r\frac{\partial T}{\partial r}\right)=\eta\dot{\gamma}^2+\frac{k}{r}\frac{\partial}{\partial r}\left(r\frac{\partial T}{\partial r}\right) \tag{5.94}$$

Chapter 6
Governing Equations for the Packing Phase

In this Chapter we establish the governing equations for the packing phase. As for the filling phase, we start with the continuity, momentum and energy equations derived in Chapter 4.

For the packing phase, it is essential to include the effect of melt compressibility. Therefore the relation $\nabla \cdot \vec{v} = 0$ may not be used to simplify the equations. We do however use all the other assumptions regarding material behaviour and geometry that were used in the previous Chapter.

6.1 The Momentum Equation for the Cavity

We begin with the momentum equation:

$$\frac{\partial}{\partial t}(\rho \vec{v}) = \rho \vec{g} + [\nabla \cdot \underline{\sigma}] - [\nabla \cdot \rho \vec{v}\vec{v}] \tag{6.1}$$

The last term may be expanded as follows:

$$[\nabla \cdot \rho \vec{v}\vec{v}] = \rho[\vec{v} \cdot \nabla \vec{v}] + (\nabla \cdot \rho \vec{v})\vec{v} \tag{6.2}$$

The LHS of Eq. (6.1) may be expanded to give

$$\begin{aligned}\frac{\partial}{\partial t}(\rho\vec{v}) &= \frac{\partial \rho}{\partial t}\vec{v} + \rho\frac{\partial \vec{v}}{\partial t}\\ &= -(\nabla\cdot\rho\vec{v})\vec{v} + \rho\frac{\partial \vec{v}}{\partial t}\end{aligned} \tag{6.3}$$

where we have used the continuity equation (see Eq. (5.1)):

$$\frac{\partial \rho}{\partial t} = -(\nabla\cdot\rho\vec{v}) \tag{6.4}$$

Rearranging Eq. (6.2) for $(\nabla\cdot\rho\vec{v})\vec{v}$ and substituting in Eq. (6.3) we obtain

$$\frac{\partial}{\partial t}(\rho\vec{v}) = \rho[\vec{v}\cdot\nabla\vec{v}] - [\nabla\cdot\rho\vec{v}\vec{v}] + \rho\frac{\partial \vec{v}}{\partial t} \tag{6.5}$$

Substituting the above equation in Eq. (6.1) we have:

$$\rho[\vec{v}\cdot\nabla\vec{v}] - [\nabla\cdot\rho\vec{v}\vec{v}] + \rho\frac{\partial \vec{v}}{\partial t} = \rho\vec{g} + [\nabla\cdot\underline{\sigma}] - [\nabla\cdot\rho\vec{v}\vec{v}] \tag{6.6}$$

That is,

$$\rho\frac{\partial \vec{v}}{\partial t} = \rho\vec{g} + [\nabla\cdot\underline{\sigma}] - \rho[\vec{v}\cdot\nabla\vec{v}] \tag{6.7}$$

Using the assumption that the melt may be considered as a generalized Newtonian fluid, we have from Section 5.1.2:

$$[\nabla\cdot\underline{\sigma}] = -\nabla p + [\nabla\cdot\eta\underline{\dot{\gamma}}] \tag{6.8}$$

Hence, equation (6.7) becomes:

$$\rho\frac{\partial \vec{v}}{\partial t} = \rho\vec{g} - \nabla p + [\nabla\cdot\eta\underline{\dot{\gamma}}] - \rho[\vec{v}\cdot\nabla\vec{v}] \tag{6.9}$$

But Eq. (6.9) is identical to Eq. (5.11) of Section 5.1.4. That is, we obtain the same momentum equation whether or not we assume the material to be compressible. Moreover, the remaining assumptions of Chapter 5 regarding the momentum equation are still valid in the packing phase and so the conclusions of that Chapter also hold. In the next section we briefly retrace the effect of these assumptions on the momentum equation. For details refer to Chapter 5.

6.1.1 Effect of Assumptions on Momentum Equation

Writing Eq. (6.8) in Cartersian coordinates and using dimensional analysis (see Section 5.2.3) to assess the relative importance of each term, we obtain the following equations:

$$\begin{aligned} \frac{\partial p}{\partial x} &= \frac{\partial}{\partial z}\left(\eta\frac{\partial v_x}{\partial z}\right) \\ \frac{\partial p}{\partial y} &= \frac{\partial}{\partial z}\left(\eta\frac{\partial v_y}{\partial z}\right) \\ \frac{\partial p}{\partial z} &= 0 \end{aligned} \tag{6.10}$$

These equations may be integrated (see Section 5.3.1) twice with respect to z to give:

$$v_x(z) = \frac{\partial p}{\partial x}\left[\int_{h^-}^{z}\frac{z'}{\eta}dz' - C(x,y)\int_{h^-}^{z}\frac{dz'}{\eta}\right] \tag{6.11}$$

$$v_y(z) = \frac{\partial p}{\partial y}\left[\int_{h^-}^{z}\frac{z'}{\eta}dz' - C(x,y)\int_{h^-}^{z}\frac{dz'}{\eta}\right] \tag{6.12}$$

Average velocities $\overline{v}_x$ and $\overline{v}_y$ may then be calculated to be:

$$\overline{v}_x = \frac{-2S_2}{h^+ - h^-}\frac{\partial p}{\partial x} \tag{6.13}$$

$$\overline{v}_y = \frac{-2S_2}{h^+ - h^-}\frac{\partial p}{\partial y} \tag{6.14}$$

where

$$S_2 = \frac{1}{2}\left\{\int_{h^-}^{h^+}\frac{z'^2}{\eta}dz' - \frac{\left(\int_{h^-}^{h^+}\frac{z'}{\eta}dz'\right)^2}{\int_{h^-}^{h^+}\frac{dz'}{\eta}}\right\} \tag{6.15}$$

Finally, enforcing the assumption of symmetry about the cavity centre-line (see Section 5.4) gives:

$$S_2 = \int_0^h \frac{z'^2}{\eta} dz' \tag{6.16}$$

6.2 The Continuity Equation for the Cavity

We now turn to the continuity equation derived in Chapter 4. The equation is

$$\frac{\partial \rho}{\partial t} + (\nabla \cdot \rho \vec{v}) = 0 \tag{6.17}$$

Expanding the second term in Eq. (6.17) we have,

$$\frac{\partial \rho}{\partial t} + \rho(\nabla \cdot \vec{v}) + \vec{v} \cdot \nabla \rho = 0 \tag{6.18}$$

Adopting a Cartesian coordinate system, this equation becomes:

$$\frac{\partial \rho}{\partial t} = -\rho\left(\frac{\partial v_x}{\partial x} + \frac{\partial v_y}{\partial y} + \frac{\partial v_z}{\partial z}\right) - \left(v_x \frac{\partial \rho}{\partial x} + v_y \frac{\partial \rho}{\partial y} + v_z \frac{\partial \rho}{\partial z}\right) \tag{6.19}$$

The density ρ is a state variable and so depends on pressure p and temperature T. Hence using the chain rule for differentiation we obtain:

$$\begin{aligned}
\frac{\partial \rho}{\partial x} &= \left(\frac{\partial \rho}{\partial p}\right)_T \frac{\partial p}{\partial x} + \left(\frac{\partial \rho}{\partial T}\right)_p \frac{\partial T}{\partial x} \\
\frac{\partial \rho}{\partial y} &= \left(\frac{\partial \rho}{\partial p}\right)_T \frac{\partial p}{\partial y} + \left(\frac{\partial \rho}{\partial T}\right)_p \frac{\partial T}{\partial y} \\
\frac{\partial \rho}{\partial z} &= \left(\frac{\partial \rho}{\partial p}\right)_T \frac{\partial p}{\partial z} + \left(\frac{\partial \rho}{\partial T}\right)_p \frac{\partial T}{\partial z} \\
\frac{\partial \rho}{\partial t} &= \left(\frac{\partial \rho}{\partial p}\right)_T \frac{\partial p}{\partial t} + \left(\frac{\partial \rho}{\partial T}\right)_p \frac{\partial T}{\partial t}
\end{aligned} \tag{6.20}$$

Substituting the above expressions in Eq. (6.19) and rearranging gives:

$$
\begin{aligned}
0 = & \frac{1}{\rho}\frac{\partial \rho}{\partial p}\left(\frac{\partial p}{\partial t} + v_x\frac{\partial p}{\partial x} + v_y\frac{\partial p}{\partial y} + v_z\frac{\partial p}{\partial z}\right) \\
& + \frac{1}{\rho}\frac{\partial \rho}{\partial T}\left(\frac{\partial T}{\partial t} + v_x\frac{\partial T}{\partial x} + v_y\frac{\partial T}{\partial y} + v_z\frac{\partial T}{\partial z}\right) + \left(\frac{\partial v_x}{\partial x} + \frac{\partial v_y}{\partial y} + \frac{\partial v_z}{\partial z}\right)
\end{aligned}
\qquad \textbf{(6.21)}
$$

Recall that density is the inverse of specific volume $\widehat{V}$. Hence we can write

$$
\begin{aligned}
\frac{1}{\rho}\frac{\partial \rho}{\partial p} &= \frac{1}{\rho}\frac{\partial}{\partial p}\left(\frac{1}{\widehat{V}}\right) \\
&= \frac{1}{\rho}\frac{\partial}{\partial \widehat{V}}\left(\frac{1}{\widehat{V}}\right)\frac{\partial \widehat{V}}{\partial p} \\
&= \frac{1}{\rho}\left(-\frac{1}{\widehat{V}^2}\right)\frac{\partial \widehat{V}}{\partial p} \\
&= -\frac{1}{\widehat{V}}\frac{\partial \widehat{V}}{\partial p} \\
&= \kappa
\end{aligned}
\qquad \textbf{(6.22)}
$$

where κ is the isothermal compressibility coefficient of the material [SS].

Similarly,

$$
\begin{aligned}
\frac{1}{\rho}\frac{\partial \rho}{\partial T} &= \frac{1}{\rho}\frac{\partial}{\partial T}\left(\frac{1}{\widehat{V}}\right) \\
&= \frac{1}{\rho}\frac{\partial}{\partial \widehat{V}}\left(\frac{1}{\widehat{V}}\right)\frac{\partial \widehat{V}}{\partial T} \\
&= \frac{1}{\rho}\left(-\frac{1}{\widehat{V}^2}\right)\frac{\partial \widehat{V}}{\partial T} \\
&= -\frac{1}{\widehat{V}}\frac{\partial \widehat{V}}{\partial T} \\
&= -\beta
\end{aligned}
\qquad \textbf{(6.23)}
$$

where β is the expansivity of the material [SS]. Both β and κ were defined by equations (2.26) and (2.23) respectively in Section 2.8.1.

Using the above, Eq. (6.21) becomes

$$
\begin{aligned}
0 = \kappa\left(\frac{\partial p}{\partial t} + v_x\frac{\partial p}{\partial x} + v_y\frac{\partial p}{\partial y} + v_z\frac{\partial p}{\partial z}\right) \\
- \beta\left(\frac{\partial T}{\partial t} + v_x\frac{\partial T}{\partial x} + v_y\frac{\partial T}{\partial y} + v_z\frac{\partial T}{\partial z}\right) + \left(\frac{\partial v_x}{\partial x} + \frac{\partial v_y}{\partial y} + \frac{\partial v_z}{\partial z}\right)
\end{aligned}
\quad \textbf{(6.24)}
$$

Because in the packing phase the cavity is volumetrically filled, the application of packing pressure tends to pressurize the melt hydrostatically. It follows that pressure gradients $\partial p/\partial x$ and $\partial p/\partial y$ are very much lower than during fill. From Eq. (6.10) we also know that $\partial p/\partial z = 0$. Given these low pressure gradients, equations (6.11) and (6.12) predict low velocities during packing. Consequently, we make the following assumption:

Assumption: Pressure convection terms may be ignored in the packing phase.

With this assumption, Eq. (6.24) becomes:

$$
\begin{aligned}
0 = \kappa\frac{\partial p}{\partial t} - \beta\left(\frac{\partial T}{\partial t} + v_x\frac{\partial T}{\partial x} + v_y\frac{\partial T}{\partial y} + v_z\frac{\partial T}{\partial z}\right) \\
+ \left(\frac{\partial v_x}{\partial x} + \frac{\partial v_y}{\partial y} + \frac{\partial v_z}{\partial z}\right)
\end{aligned}
\quad \textbf{(6.25)}
$$

6.3 The Energy Equation for the Cavity

The energy equation was derived in Chapter 4 and is repeated below for convenience:

$$
\rho c_p\left(\frac{\partial T}{\partial t} + \vec{v}\cdot\nabla T\right) = \beta T\left(\frac{\partial p}{\partial t} + \vec{v}\cdot\nabla p\right) + p\nabla\cdot\vec{v} + (\underline{\sigma} : \{\nabla\vec{v}\}) + \nabla\cdot(k\nabla T) \quad \textbf{(6.26)}
$$

In the packing phase we use the same assumptions regarding material behaviour as used in the filling phase (see Section 5.1) except that the fluid is now considered compressible. Assuming constant thermal conductivity, the energy equation becomes:

$$
\rho c_p\left(\frac{\partial T}{\partial t} + \vec{v}\cdot\nabla T\right) = \beta T\left(\frac{\partial p}{\partial t} + \vec{v}\cdot\nabla p\right) + p\nabla\cdot\vec{v} + (\underline{\sigma} : \{\nabla\vec{v}\}) + k\nabla^2 T \quad \textbf{(6.27)}
$$

We now turn our attention to the term involving the stress tensor: $\underline{\sigma} : \{\nabla\vec{v}\}$.

In Section 3.1.3 it was shown that the stress tensor may always be written in the form $\underline{\sigma} = -p\underline{I} + \underline{\tau}$. Where $\underline{\tau}$ is defined as follows:

$$\begin{aligned}\underline{\tau} &= \eta\underline{\dot{\gamma}}' \\ &= \eta\left(\underline{\dot{\gamma}} - \frac{2}{3}(\nabla\cdot\vec{v})\underline{I}\right) \\ &= \eta\left(\nabla\vec{v} + \nabla\vec{v}^{\dagger} - \frac{2}{3}(\nabla\cdot\vec{v})\underline{I}\right)\end{aligned} \tag{6.28}$$

Note that because the fluid is compressible, we are using the deviatoric rate of strain tensor defined earlier in Section 3.4. Hence the shear rate is, from Eq. (3.21), given by

$$\begin{aligned}\dot{\gamma}'^2 &= \frac{1}{2}\underline{\dot{\gamma}}' : \underline{\dot{\gamma}}' \\ &= \frac{1}{2}\left(\underline{\dot{\gamma}} - \frac{2}{3}(\nabla\cdot\vec{v})\underline{I}\right) : \left(\underline{\dot{\gamma}} - \frac{2}{3}(\nabla\cdot\vec{v})\underline{I}\right) \\ &= \frac{1}{2}\left[(\underline{\dot{\gamma}} : \underline{\dot{\gamma}}) - \frac{4}{3}(\nabla\cdot\vec{v}) : \underline{I} + \frac{4}{9}(\nabla\cdot\vec{v})^2\underline{I} : \underline{I}\right]\dot{\gamma} \\ &= \frac{1}{2}\left[(\underline{\dot{\gamma}} : \underline{\dot{\gamma}}) - \frac{8}{3}(\nabla\cdot\vec{v})^2 + \frac{4}{3}(\nabla\cdot\vec{v})^2\right] \\ &= \frac{1}{2}\left[(\underline{\dot{\gamma}} : \underline{\dot{\gamma}}) - \frac{4}{3}(\nabla\cdot\vec{v})^2\right] \\ &= \frac{1}{2}\left[(\nabla\vec{v} + \nabla\vec{v}^{\dagger}) : (\nabla\vec{v} + \nabla\vec{v}^{\dagger}) - \frac{4}{3}(\nabla\cdot\vec{v})^2\right] \\ &= (\nabla\vec{v} : \nabla\vec{v}) + (\nabla\vec{v}^{\dagger} : \nabla\vec{v}) - \frac{2}{3}(\nabla\cdot\vec{v})^2\end{aligned} \tag{6.29}$$

Using this result, we obtain,

$$
\begin{aligned}
(\underline{\sigma} : \{\nabla\vec{v}\}) &= (-p\underline{I} + \underline{\tau}) : \nabla\vec{v} \\
&= (\{-p\underline{I} + \eta\underline{\dot{\gamma}}'\} : \nabla\vec{v}) \\
&= (-p\underline{I} : \nabla\vec{v}) + (\eta\underline{\dot{\gamma}}' : \nabla\vec{v}) \\
&= -p(\nabla\cdot\vec{v}) + \left(\eta\left\{\nabla\vec{v} + \nabla\vec{v}^{\dagger} - \frac{2}{3}(\nabla\cdot\vec{v})\underline{I}\right\} : \nabla\vec{v}\right) \\
&= -p(\nabla\cdot\vec{v}) + \eta\left(\nabla\vec{v} : \nabla\vec{v} + \nabla\vec{v}^{\dagger} : \nabla\vec{v} - \frac{2}{3}(\nabla\cdot\vec{v})\underline{I} : \nabla\vec{v}\right) \\
&= -p(\nabla\cdot\vec{v}) + \eta\left(\nabla\vec{v} : \nabla\vec{v} + \nabla\vec{v}^{\dagger} : \nabla\vec{v} - \frac{2}{3}(\nabla\cdot\vec{v})^2\right) \\
&= -p(\nabla\cdot\vec{v}) + \eta\dot{\gamma}'^2
\end{aligned} \tag{6.30}
$$

Substituting the above expression in Eq. (6.27 the energy equation for packing takes the form:

$$
\rho c_p\left(\frac{\partial T}{\partial t} + \vec{v}\cdot\nabla T\right) = \beta T\left(\frac{\partial p}{\partial t} + \vec{v}\cdot\nabla p\right) + \eta\dot{\gamma}'^2 + k\nabla^2 T \tag{6.31}
$$

In Cartesian coordinates, the energy equation takes the form:

$$
\begin{aligned}
\rho c_p\left(\frac{\partial T}{\partial t} + v_x\frac{\partial T}{\partial x} + v_y\frac{\partial T}{\partial y} + v_z\frac{\partial T}{\partial z}\right) &= \beta T\left(\frac{\partial p}{\partial t} + v_x\frac{\partial p}{\partial x} + v_y\frac{\partial p}{\partial y} + v_z\frac{\partial p}{\partial z}\right) \\
&\quad + \eta\dot{\gamma}'^2 + k\left(\frac{\partial^2 T}{\partial x^2} + \frac{\partial^2 T}{\partial y} + \frac{\partial^2 T}{\partial z^2}\right)
\end{aligned} \tag{6.32}
$$

Estimating the order of magnitude of each of these terms as in Section 5.2.4 leads us to ignore pressure, convection and conduction in the x and y directions. In contrast to the filling phase, we retain the term involving $\partial p/\partial t$. This is because although the pressure gradient terms $\partial p/\partial x$, $\partial p/\partial y$ and $\partial p/\partial z$ are relatively low in the packing phase, $\partial p/\partial t$ may be quite large. This is particularly true at the instant at which the cavity is volumetrically filled. At this time the pressure throughout the cavity increases suddenly as the melt tends towards a fluid under hydrostatic pressure. The energy equation therefore becomes:

$$
\rho c_p\left(\frac{\partial T}{\partial t} + v_x\frac{\partial T}{\partial x} + v_y\frac{\partial T}{\partial y} + v_z\frac{\partial T}{\partial z}\right) = \beta T\frac{\partial p}{\partial t} + \eta\dot{\gamma}'^2 + k\frac{\partial^2 T}{\partial z^2} \tag{6.33}
$$

Finally, we simplify this further by ignoring convection in the z-direction. The energy equation to be used in the packing phase is therefore:

$$\rho c_p \left(\frac{\partial T}{\partial t} + v_x \frac{\partial T}{\partial x} + v_y \frac{\partial T}{\partial y} \right) = \beta T \frac{\partial p}{\partial t} + \eta \dot{\gamma}'^2 + k \frac{\partial^2 T}{\partial z^2} \tag{6.34}$$

To complete this section we consider the term $\dot{\gamma}'^2$. Using dimensional analysis it can be shown that, in Cartesian coordinates, and flow in the cavity

$$\begin{aligned} \dot{\gamma}'^2 &= \sqrt{\left(\frac{\partial v_x}{\partial z} \right)^2 + \left(\frac{\partial v_y}{\partial z} \right)^2} \\ &= \dot{\gamma}^2 \end{aligned} \tag{6.35}$$

That is, the shear rate for the compressible case in packing is for practical purposes the same as for the filling phase. We therefore drop the prime on the shear rate symbol in what follows.

6.4 Integration of the Continuity Equation

The energy equation (6.33) may be rearranged to give:

$$\frac{\partial T}{\partial t} + v_x \frac{\partial T}{\partial x} + v_y \frac{\partial T}{\partial y} + v_z \frac{\partial T}{\partial z} = \frac{1}{\rho c_p} \left(\beta T \frac{\partial p}{\partial t} + \eta \dot{\gamma}^2 + k \frac{\partial^2 T}{\partial z^2} \right) \tag{6.36}$$

This may be substituted in the continuity equation (6.25) to give:

$$\begin{aligned} 0 = \kappa \frac{\partial p}{\partial t} - \frac{\beta}{\rho c_p} \left(\beta T \frac{\partial p}{\partial t} + \eta \dot{\gamma}^2 + k \frac{\partial^2 T}{\partial z^2} \right) \\ + \left(\frac{\partial v_x}{\partial x} + \frac{\partial v_y}{\partial y} + \frac{\partial v_z}{\partial z} \right) \end{aligned} \tag{6.37}$$

Equation (6.37) may be integrated with respect to z over the cavity thickness to give:

$$
\begin{aligned}
0 &= \int_{H^-}^{H^+} \kappa \frac{\partial p}{\partial t} dz - \int_{H^-}^{H^+} \frac{\beta}{\rho c_p} \left(\beta T \frac{\partial p}{\partial t} + \eta \dot{\gamma}^2 + k \frac{\partial^2 T}{\partial z^2} \right) dz \\
&\quad + \int_{H^-}^{H^+} \left(\frac{\partial v_x}{\partial x} + \frac{\partial v_y}{\partial y} + \frac{\partial v_z}{\partial z} \right) dz \\
&= \int_{H^-}^{H^+} \left(\kappa - \frac{\beta^2 T}{\rho c_p} \right) \frac{\partial p}{\partial t} dz - \int_{H^-}^{H^+} \frac{\beta}{\rho c_p} \left(\eta \dot{\gamma}^2 + k \frac{\partial^2 T}{\partial z^2} \right) dz \\
&\quad + \int_{h^-}^{h^+} \left(\frac{\partial v_x}{\partial x} + \frac{\partial v_y}{\partial y} + \frac{\partial v_z}{\partial z} \right) dz
\end{aligned}
\tag{6.38}
$$

Note that the last integral in the above expression is identical to that evaluated in Section 5.3.2 and so, using the results of that section (Eq. (5.57) we obtain:

$$
\begin{aligned}
0 = \int_{H^-}^{H^+} \left(\kappa - \frac{\beta^2 T}{\rho c_p} \right) \frac{\partial p}{\partial t} dz - \int_{H^-}^{H^+} \frac{\beta}{\rho c_p} \left(\eta \dot{\gamma}^2 + k \frac{\partial^2 T}{\partial z^2} \right) dz \\
- 2 \frac{\partial}{\partial x} \left(S_2 \frac{\partial p}{\partial x} \right) - 2 \frac{\partial}{\partial y} \left(S_2 \frac{\partial p}{\partial y} \right)
\end{aligned}
\tag{6.39}
$$

where

$$
S_2 = \frac{1}{2} \left\{ \int_{h^-}^{h^+} \frac{z^2}{\eta} dz - \frac{\left(\int_{h^-}^{h^+} \frac{z}{\eta} dz \right)^2}{\int_{h^-}^{h^+} \frac{dz}{\eta}} \right\}
\tag{6.40}
$$

In order to simplify this expression we make the same assumption (see Section 5.4) as in the filling phase regarding symmetry of the flow field;

> **Assumption:** The flow field is assumed to be symmetrical about the cavity centre line.

With this assumption, Eq. (6.39) has the same form but the limits of integration can be changed and S_2 is simplified. That is, we obtain:

$$0 = \int_0^H \left(\kappa - \frac{\beta^2 T}{\rho c_p}\right)\frac{\partial p}{\partial t}dz - \int_0^H \frac{\beta}{\rho c_p}\left(\eta\dot{\gamma}^2 + k\frac{\partial^2 T}{\partial z^2}\right)dz - \frac{\partial}{\partial x}\left(S_2\frac{\partial p}{\partial x}\right) - \frac{\partial}{\partial y}\left(S_2\frac{\partial p}{\partial y}\right) \tag{6.41}$$

where

$$S_2 = \int_0^h \frac{z^2}{\eta}dz \tag{6.42}$$

For typical values of κ, β, ρ and c_p the term $\beta^2 T/\rho c_p$is much smaller than κ and so it is ignored.

We now define an average value of compressibility coefficient, $\overline{\kappa}$:

$$\overline{\kappa} = \frac{1}{H}\int_0^H \kappa\, dz \tag{6.43}$$

Noting that $\partial p/\partial t$ is independent of z, Equation (6.41) becomes:

$$0 = \overline{\kappa}H\frac{\partial p}{\partial t} - \int_0^H \frac{\beta}{\rho c_p}\left(\eta\dot{\gamma}^2 + k\frac{\partial^2 T}{\partial z^2}\right)dz - \frac{\partial}{\partial x}\left(S_2\frac{\partial p}{\partial x}\right) - \frac{\partial}{\partial y}\left(S_2\frac{\partial p}{\partial y}\right) \tag{6.44}$$

6.5 Packing Equations for Runners

For the packing phase, the same material assumptions given in Section 5.1 are used except that the material is now taken to be compressible.

The governing equations are therefore:

Continuity

$$\frac{\partial \rho}{\partial t} + (\nabla \cdot \rho \vec{v}) = 0 \tag{6.45}$$

Momentum: (see Eq. (6.9)

$$\rho \frac{\partial \vec{v}}{\partial t} = \rho \vec{g} - \nabla p + [\nabla \cdot \eta \underline{\dot{\gamma}}] - \rho [\vec{v} \cdot \nabla \vec{v}] \tag{6.46}$$

Energy

$$\rho c_p \left(\frac{\partial T}{\partial t} + \vec{v} \cdot \nabla T \right) = \beta T \left(\frac{\partial p}{\partial t} + \vec{v} \cdot \nabla p \right) + \eta \dot{\gamma}^2 + k \nabla^2 T \tag{6.47}$$

Each of these will be considered for runners in the following sections.

6.5.1 Continuity

The continuity equation (6.45), may be expressed in cylindrical coordinates using the relationships given in Appendix B as follows:

$$\begin{aligned} \frac{\partial \rho}{\partial t} + \rho \left(\frac{1}{r} \frac{\partial}{\partial r} (r v_r) + \frac{1}{r} \frac{\partial v_\theta}{\partial \theta} + \frac{\partial v_x}{\partial x} \right) \\ + v_r \frac{\partial \rho}{\partial r} + \frac{v_\theta}{r} \frac{\partial \rho}{\partial \theta} + v_x \frac{\partial \rho}{\partial x} = 0 \end{aligned} \tag{6.48}$$

Due to symmetry about the axis of the runner, terms involving derivatives with respect to θ are zero and the continuity equation becomes:

$$\begin{aligned} \frac{\partial \rho}{\partial t} + \rho \left(\frac{1}{r} \frac{\partial}{\partial r} (r v_r) + \frac{\partial v_x}{\partial x} \right) \\ + v_r \frac{\partial \rho}{\partial r} + v_x \frac{\partial \rho}{\partial x} = 0 \end{aligned} \tag{6.49}$$

The density ρ depends on both the temperature and the pressure. Therefore, using the chain rule for differentiation, the definitions of expansivity, β, and compressibility κ given in Section 2.8.1 and equations (6.22) and (6.23) we have:

$$\begin{aligned}
\frac{\partial \rho}{\partial t} &= \left(\frac{\partial \rho}{\partial p}\right)_T \frac{\partial p}{\partial t} + \left(\frac{\partial \rho}{\partial T}\right)_p \frac{\partial T}{\partial t} \\
&= \rho\kappa \frac{\partial p}{\partial t} - \rho\beta \frac{\partial T}{\partial t} \\
\frac{\partial \rho}{\partial r} &= \left(\frac{\partial \rho}{\partial p}\right)_T \frac{\partial p}{\partial r} + \left(\frac{\partial \rho}{\partial T}\right)_p \frac{\partial T}{\partial r} \\
&= \rho\kappa \frac{\partial p}{\partial r} - \rho\beta \frac{\partial T}{\partial r} \\
\frac{\partial \rho}{\partial x} &= \left(\frac{\partial \rho}{\partial p}\right)_T \frac{\partial p}{\partial x} + \left(\frac{\partial \rho}{\partial T}\right)_p \frac{\partial T}{\partial x} \\
&= \rho\kappa \frac{\partial p}{\partial x} - \rho\beta \frac{\partial T}{\partial x}
\end{aligned} \tag{6.50}$$

Substituting these expressions in (6.49) and rearranging gives

$$\begin{aligned}
\kappa\left(\frac{\partial p}{\partial t} + v_r \frac{\partial p}{\partial r} + v_x \frac{\partial p}{\partial x}\right) - \beta\left(\frac{\partial T}{\partial t} + v_r \frac{\partial T}{\partial r} + v_x \frac{\partial T}{\partial x}\right) \\
+ \frac{1}{r}\frac{\partial}{\partial r}(r v_r) + \frac{\partial v_x}{\partial x} = 0
\end{aligned} \tag{6.51}$$

Using the same reasoning as in Section 6.2, we assume that the pressure convection terms may be ignored and so finally obtain,

$$\kappa \frac{\partial p}{\partial t} - \beta\left(\frac{\partial T}{\partial t} + v_r \frac{\partial T}{\partial r} + v_x \frac{\partial T}{\partial x}\right) + \frac{1}{r}\frac{\partial}{\partial r}(r v_r) + \frac{\partial v_x}{\partial x} = 0 \tag{6.52}$$

6.5.2 Momentum:

In Section 6.1 it was shown that the momentum equations for the cavity are identical for compressible and incompressible fluids. This is also true for flow in runners and so the momentum equations in the packing phase are the same as those used for the filling phase. They are, (from Section 5.5.2),

$$\frac{\partial p}{\partial r} = 0$$
$$\frac{\partial p}{\partial \theta} = 0 \qquad (6.53)$$
$$\frac{\partial p}{\partial x} = \frac{1}{r}\frac{\partial}{\partial r}\left(r\eta\frac{\partial v_x}{\partial r}\right)$$

6.5.3 Energy

The energy equation in cylindrical coordinates is (see Section 5.5.3).

$$\rho c_p\left(\frac{\partial T}{\partial t} + v_r\frac{\partial T}{\partial r} + v_x\frac{\partial T}{\partial x}\right) = \beta T\left(\frac{\partial p}{\partial t} + v_r\frac{\partial p}{\partial r} + v_x\frac{\partial p}{\partial x}\right) + \eta\dot{\gamma}^2 + k\left(\frac{1}{r}\frac{\partial}{\partial r}\left(r\frac{\partial T}{\partial r}\right) + \frac{\partial^2 T}{\partial x^2}\right) \qquad (6.54)$$

Ignoring low order terms and retaining the time rate of pressure term, as done in the cavity (see Section 6.3), this equation simplifies to:

$$\rho c_p\left(\frac{\partial T}{\partial t} + v_r\frac{\partial T}{\partial r} + v_x\frac{\partial T}{\partial x}\right) = \beta T\frac{\partial p}{\partial t} + \eta\dot{\gamma}^2 + \frac{k}{r}\frac{\partial}{\partial r}\left(r\frac{\partial T}{\partial r}\right) \qquad (6.55)$$

where

$$\dot{\gamma} = \frac{\partial v_x}{\partial r} \qquad (6.56)$$

Finally, ignoring convection in the radial direction, we obtain the energy equation to be used for thermal calculations in the packing phase:

$$\rho c_p\left(\frac{\partial T}{\partial t} + v_x\frac{\partial T}{\partial x}\right) = \beta T\frac{\partial p}{\partial t} + \eta\dot{\gamma}^2 + \frac{k}{r}\frac{\partial}{\partial r}\left(r\frac{\partial T}{\partial r}\right) \qquad (6.57)$$

6.6 Integration of the Continuity Equation for Runners

The procedure below is identical to that for the cavity (see Section 6.4). First the energy equation (6.55) is rearranged to give:

$$\frac{\partial T}{\partial t} + v_r\frac{\partial T}{\partial r} + v_x\frac{\partial T}{\partial x} = \frac{1}{\rho c_p}\left\{\beta T\frac{\partial p}{\partial t} + \eta\dot{\gamma}^2 + \frac{k}{r}\frac{\partial}{\partial r}\left(r\frac{\partial T}{\partial r}\right)\right\} \qquad (6.58)$$

Substituting this in the continuity equation (6.52) we obtain,

$$
\begin{aligned}
0 &= \kappa\frac{\partial p}{\partial t} - \beta\left(\frac{\partial T}{\partial t} + v_r\frac{\partial T}{\partial r} + v_x\frac{\partial T}{\partial x}\right) \\
&\quad + \frac{1}{r}\frac{\partial}{\partial r}(rv_r) + \frac{\partial v_x}{\partial x} \\
&= \kappa\frac{\partial p}{\partial t} - \frac{\beta}{\rho c_p}\left\{\beta T\frac{\partial p}{\partial t} + \eta\dot{\gamma}^2 + \frac{k}{r}\frac{\partial}{\partial r}\left(r\frac{\partial T}{\partial r}\right)\right\} \\
&\quad + \frac{1}{r}\frac{\partial}{\partial r}(rv_r) + \frac{\partial v_x}{\partial x}
\end{aligned}
\tag{6.59}
$$

The quantity $\beta^2 T/\rho c_p$ is much smaller than κ and so may be ignored. Hence (6.59) becomes,

$$
\begin{aligned}
0 = \kappa\frac{\partial p}{\partial t} &- \frac{\beta}{\rho c_p}\left\{\eta\dot{\gamma}^2 + \frac{k}{r}\frac{\partial}{\partial r}\left(r\frac{\partial T}{\partial r}\right)\right\} \\
&+ \frac{1}{r}\frac{\partial}{\partial r}(rv_r) + \frac{\partial v_x}{\partial x}
\end{aligned}
\tag{6.60}
$$

Taking A_c as the area of the runner cross-section, we define an average value of compressibility coefficient, $\overline{\kappa}_c$ as follows

$$
\begin{aligned}
\overline{\kappa}_c &= \frac{1}{A_c}\int\limits_{A_c} \kappa\, dA \\
&= \frac{2}{R^2}\int\limits_0^R \kappa r\, dr
\end{aligned}
\tag{6.61}
$$

Noting that dp/dt is independent of z and using the average compressibility coefficient, Eq. (6.60) can be integrated over the cross-section of the runner to give:

$$
\begin{aligned}
0 &= 2\pi \int_0^R \left[\kappa \frac{\partial p}{\partial t} - \frac{\beta}{\rho c_p} \left\{ \eta \dot{\gamma}^2 + \frac{k}{r} \frac{\partial}{\partial r} \left(r \frac{\partial T}{\partial r} \right) \right\} \right. \\
&\quad \left. + \frac{1}{r} \frac{\partial}{\partial r} (r v_r) + \frac{\partial v_x}{\partial x} \right] r dr \\
&= 2\pi \left[\frac{\partial p}{\partial t} \int_0^R \kappa r dr - \int_0^R \frac{\beta}{\rho c_p} \left\{ \eta \dot{\gamma}^2 + \frac{k}{r} \frac{\partial}{\partial r} \left(r \frac{\partial T}{\partial r} \right) \right\} r dr \right. \\
&\quad \left. + \int_0^R \left\{ \frac{1}{r} \frac{\partial}{\partial r} (r v_r) + \frac{\partial v_x}{\partial x} \right\} r dr \right] \\
&= 2\pi \left[\frac{1}{2} \overline{\kappa}_c R^2 \frac{\partial p}{\partial t} - \int_0^R \frac{\beta}{\rho c_p} \left\{ \eta \dot{\gamma}^2 + \frac{k}{r} \frac{\partial}{\partial r} \left(r \frac{\partial T}{\partial r} \right) \right\} r dr \right. \\
&\quad \left. + \int_0^{r^+} \left\{ \frac{1}{r} \frac{\partial}{\partial r} (r v_r) + \frac{\partial v_x}{\partial x} \right\} r dr \right]
\end{aligned}
\tag{6.62}
$$

The last integral on the RHS of the above equation has already been considered in the filling phase (see Section 5.5.5), where we found (Eq. (5.89)),

$$
2\pi \int_0^{r^+} \left\{ \frac{1}{r} \frac{\partial}{\partial r} (r v_r) + \frac{\partial v_x}{\partial x} \right\} r dr = -\pi \frac{\partial}{\partial x} \left(r^+ S_1 \frac{\partial p}{\partial x} \right) \tag{6.63}
$$

where

$$
S_1 = \frac{1}{2r^+} \int_0^{r^+} \frac{r^3}{\eta} dr \tag{6.64}
$$

Hence Eq. (6.62) becomes:

$$0 = \overline{\kappa}_c R^2 \frac{\partial p}{\partial t} - 2 \int_0^R \frac{\beta}{\rho c_p} \left\{ \eta \dot{\gamma}^2 + \frac{k}{r} \frac{\partial}{\partial r} \left(r \frac{\partial T}{\partial r} \right) \right\} r dr - \frac{\partial}{\partial x} \left(r^+ S_1 \frac{\partial p}{\partial x} \right) \tag{6.65}$$

6.7 Summary

Starting with the very general equations derived in Chapter 4, a number of assumptions were made to simplify the equations. These assumptions were similar to those used for the filling phase. The most significant difference was that the fluid is assumed to be compressible for the packing phase.

The resulting equations are:

- **Pressure Equation for the Cavity:** A combination of the momentum and continuity equations results in a single equation for pressure, p:

$$0 = \overline{\kappa} H \frac{\partial p}{\partial t} - \int_0^H \frac{\beta}{\rho c_p} \left(\eta \dot{\gamma}^2 + k \frac{\partial^2 T}{\partial z^2} \right) dz - \frac{\partial}{\partial x} \left(S_2 \frac{\partial p}{\partial x} \right) - \frac{\partial}{\partial y} \left(S_2 \frac{\partial p}{\partial y} \right) \tag{6.66}$$

where $S_2 = \int_0^h \frac{z'^2}{\eta} dz'$

- **Energy Equation for Cavity:**

$$\rho c_p \left(\frac{\partial T}{\partial t} + v_x \frac{\partial T}{\partial x} + v_y \frac{\partial T}{\partial y} \right) = \beta T \frac{\partial p}{\partial t} + \eta \dot{\gamma}^2 + k \frac{\partial^2 T}{\partial z^2} \tag{6.67}$$

where $\dot{\gamma}^2 = \sqrt{(\partial v_x / \partial z)^2 + (\partial v_y / \partial z)^2}$

- **Pressure Equation for Circular Runners:**

$$0 = \overline{\kappa}_c R^2 \frac{\partial p}{\partial t} - 2 \int_0^R \frac{\beta}{\rho c_p} \left\{ \eta \dot{\gamma}^2 + \frac{k}{r} \frac{\partial}{\partial r} \left(r \frac{\partial T}{\partial r} \right) \right\} r dr - \frac{\partial}{\partial x} \left(r^+ S_1 \frac{\partial p}{\partial x} \right) \tag{6.68}$$

where $S_1 = \frac{1}{2r+} \int_0^{r+} \frac{r^3}{\eta} \, dr$

- **Energy Equation for Circular Runners:**

$$\rho c_p \left(\frac{\partial T}{\partial t} + v_x \frac{\partial T}{\partial x} \right) = \beta T \frac{\partial p}{\partial t} + \eta \dot{\gamma}^2 + \frac{k}{r} \frac{\partial}{\partial r} \left(r \frac{\partial T}{\partial r} \right) \tag{6.69}$$

where $\dot{\gamma} = \partial v_x / \partial r$

Chapter 7
Numerical Methods Used in Flow Analysis: The Finite Element Method

Moldflow flow analysis software utilizes the finite element method for numerical solution of the equations describing fluid flow.

In this chapter we give a brief introduction to the concepts of finite element analysis. No attempt will be made to present a rigorous mathematical treatment, rather, we will focus on the essential ideas required for understanding Moldflow's flow analysis software.

Our treatment is by no means complete but is included as background for later Chapters. Readers who are interested in more detail should consult some of the many books devoted to the topic [BD], [BK], [HT], [HU].

7.1 Introduction to the Finite Element Method

Generally speaking, the equations governing the flow of a polymer melt cannot be solved in any but the simplest of geometries. Since injection molds have different geometries, a general method of solution is required. The finite element method is particularly well suited to this type of problem.

The Finite Element Method (FEM) is a general method for solving engineering problems. Historically, finite element analysis (FEA) began in structural analysis. Soon, however, it was realized that the method could be applied to a very wide class of problems.

7.1.1 Basic Terminology

Before getting into details we need some terminology.

A *system* is an object and/or process under investigation. Typically the system has a physical meaning and you want to know the effect of outside influences on the system. In our case the system is the injection molding process.

A *mathematical model* is a set of governing equations that describes the actual system. Frequently a mathematical model that describes the system exactly does not exist or is too complicated to be of practical use. In these cases simplifications are made so that a mathematical model can be used. In flow analysis, we have mathematical models for material viscosity as a function of temperature and shear rate and relationships between the geometry of the mold, pressure and flow rate.

In Chapter 4 we derived the basic equations governing the flow of polymer melts. These equations were then simplified for the filling and packing phases in Chapters 5 and 6 respectively. These simplified equations form the mathematical model we will use.

A *solution domain* is a region of space throughout which the mathematical model is defined and solutions are sought. Often the solution domain also includes the period of time over which changes in the system take place. In flow analysis the solution domain is the physical space in which the plastic flows (the model of the mold cavity) and the time during which filling and packing takes place. The physical space of the solution domain is defined by the user when preparing a FEA model of the cavity geometry.

Boundary conditions are physical effects that cause the system to change. For example, in flow analysis, the flow rate of the melt, the mold wall temperature and the pressure at the flow front are boundary conditions. Generally, for a given mathematical model, there will be many suitable solutions. Boundary conditions restrict the number of possible solutions and ensure a unique solution to the problem under consideration.

7.1.2 The Finite Element Approach

A finite element analysis may be broken into the following steps:

- modeling of the solution domain
- meshing of the solution domain
- derivation of element equations
- assembly of element equations
- application of boundary conditions
- solution of system equations
- display of results

Modeling and meshing are frequently called pre-processing.

Derivation of element equations, assembly of element equations, application of boundary conditions and solution of the system equations are all performed by the flow analysis software. In later chapters we will derive the element equations and discuss the boundary conditions for flow analysis.

Results of finite element analysis are usually displayed graphically.

Modeling

Modeling is the process in which you create a representation of the domain. This representation is called a *model* and should not be confused with the mathematical models discussed above. All flow analysis software provides some modeling capability. Often the cavity geometry is described in a CAD system. Typically, the user starts the modeling process by defining points. These are then linked to form surfaces. Thicknesses may be assigned to the surfaces so as to represent the actual mold cavity.

Meshing

Meshing is the process in which the domain is discretized into smaller regions called *elements*. Meshing is accomplished by defining points on the surfaces called *nodes*. The nodes are linked to form the elements. It is the nodes and elements that constitute the *mesh*.

Flow analysis software uses one and two dimensional elements. One dimensional elements represent beams or runner systems and are defined by two nodes. Two dimensional elements are used to represent the inside of the cavity. Usually two dimensional elements are defined by three nodes and are triangular in shape. For flow analysis, a triangular mesh is adequate as any surface can be "covered" by a lattice of triangles. This allows meshing of surfaces to be done automatically. Quadrilateral elements defined by four nodes are very common in structural analysis and frequently outperform triangular elements. They are sometimes used for flow analysis but offer no advantage over triangles.

Derivation of Element Equations

Injection molded parts often have complex shapes and direct solution of the necessary mathematical models is not possible. However, by breaking up the domain into a number of smaller regions (elements), progress can be made.

Due to the simple shape and small size of elements, algebraic equations that represent the mathematical model can be derived for each element This usually entails making some approximations. Nevertheless, these are reasonable given that the element is such a small region of the domain.

Algebraically the equations are identical for a given type of element. For example, all triangular elements in the mesh will have the same algebraic equation. Only the values of variables will change from element to element.

The fact that element equations can be derived for simple regions is the basic concept behind the finite element method.

Fortunately, element equations are handled by the software. You need only specify the type of element you wish to use. All the necessary mathematical analysis has been done by the software developer.

Assembly of Element Equations

The element equations are only valid for small regions of the solution domain. To find a solution for the whole domain, the element equations are assembled to form *system equations*. This procedure is done automatically by the software. The system equations generally comprise a large set of equations that may be solved to give solutions for the whole domain.

System equations are usually represented in matrix form. For flow analysis the equations may take the form:

$$\{Q\} = [K_f]\{P\} \qquad \textbf{(7.1)}$$

where $\{Q\}$ is the nodal flow rate vector, $[K_f]$ is called the stiffness matrix and $\{P\}$ is the nodal pressure vector.

For a different application such as structural analysis,, the system equations may be written as

$$\{F\} = [K_s]\{U\} \qquad \textbf{(7.2)}$$

where $\{F\}$ is the nodal force vector, $[K_s]$ is the stiffness matrix and $\{U\}$ is the nodal displacement vector.

Note that equations (7.1) and (7.2) have identical structure. This fact highlights the versatility of the FEM. Any problem that is solved using the FEM involves system equations of this form.

Application of the Boundary Conditions

The boundary conditions specify any physical effects that cause the system to change. They are introduced by modifying the system equations. Once again this aspect of analysis is performed by the software automatically although you give the program some help. For example, in flow analysis, you specify the mold temperature, melt temperature and injection time.

Nodes are convenient places to define boundary conditions. For example in flow analysis you can define any node to be the injection point and so define a flow rate at this node.

Solution of the System Equations

System equations are solved using numerical techniques. There are two main categories: direct and iterative. An iterative method is used in flow analysis software.

Generally you do not have to worry about solution of the equations, the software does it internally. However, in most programs you can set tolerances on the accuracy of solution. This facility can be used in two ways: to increase accuracy or to decrease accuracy while speeding up the program. The latter may be useful if you are adjusting gate positions and simply want some estimates of flow front position.

Display of Results

Results from the analysis consist of files that contain numbers in tabular format. Direct interpretation of these files is virtually impossible. Consequently results are usually displayed graphically by plotting contours of the calculated physical variable over the model of the part.

7.2 The Nature of a Finite Element Solution

Consider the problem of approximating a simple curve by line segments. An example is shown in Figure (7.19).

Figure 7.19: Approximation of Simple Curve

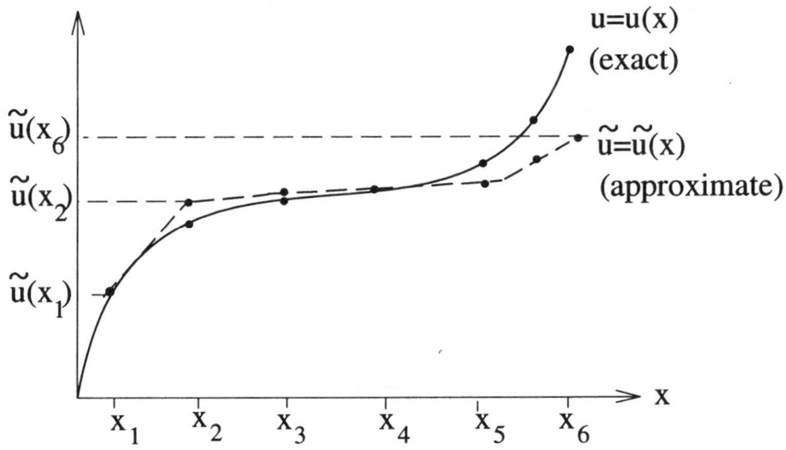

It is clear that the accuracy of the approximation depends on

1. the number of line segments used and their shape. In the figure we have used straight lines, but the accuracy could be improved if we used curved line segments.
2. the accuracy with which the points $\tilde{u}(x_i)$ approximate the exact solution.

In finite element terms, this approximation would be described as follows. The solution domain, $\{x : x_1 \leq x \leq x_6\}$ is divided into five sub-domains: $\{x : x_1 \leq x \leq x_2\}$, $\{x : x_2 \leq x \leq x_3\}$, $\{x : x_3 \leq x \leq x_4\}$, $\{x : x_4 \leq x \leq x_5\}$and $\{x : x_5 \leq x \leq x_6\}$. Each of these sub-domains is called an element. The size of each element can vary. Each element is defined by two nodes. Nodes are located at x_1, x_2, x_3, x_4, x_5 and x_6. Within an element, the approximate solution is assumed to be linear.

The finite element method gives us an approximation to the exact solution by determining the nodal values $\tilde{u}(x_1)$, $\tilde{u}(x_2)$, $\tilde{u}(x_3)$, etc. in such a way that for a given number of elements, of given size and shape, the approximate solution is "best" in some sense.

The linear element used above is the simplest type of element. It is possible to use higher order elements which approximate the solution with a series of curves. However, we will consider only linear elements.

The above example is only one-dimensional. It is possible to use finite element methods in higher dimensions also. For a two dimensional problem, such as flow through a mold cavity or temperature conduction in a plate, the elements will be patches on the domain and the approximated solution is a collection of plane surfaces. For example, Figure (7.20) shows a finite element solution to a two dimensional problem in which the elements are triangular, with linear interpolation between nodes.

Figure 7.20: Finite Element Solution of Two-Dimensional Problem

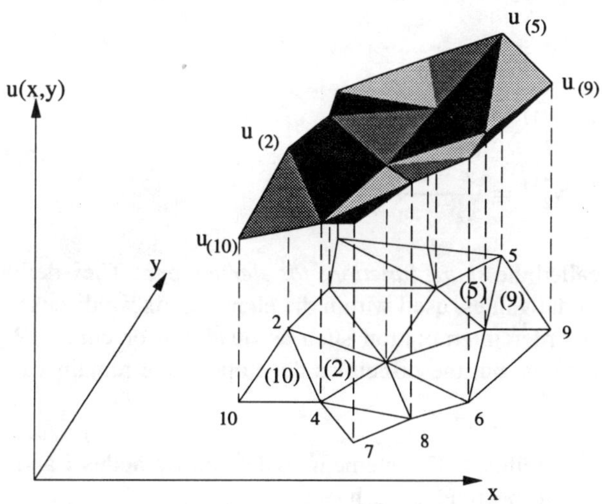

Just as in the one-dimensional case, the quality of solution is determined by the accuracy to which the nodal values approximate the exact solution and the number and shape of elements used.

7.3 Shape Functions

As pointed out in the section above, the way we interpolate the approximate solution between nodes has considerable bearing on the accuracy of the finite element method. Before discussing this further we define some notation.

When dealing with the solution to a problem with exact solution $u(x)$, we will denote the approximate solution by $\tilde{u}(x)$. Also, the nodal values of the approximate solution will be denoted by u_i, where i is the node number.

Consider element 1 in the example of Figure (7.19). We will denote the approximate solution in element 1 by $\tilde{u}^{(1)}(x)$. This is a straight line and may be defined in terms of the nodal values u_1, u_2 as follows:

$$\begin{aligned}\tilde{u}^{(1)}(x) &= \left(\frac{x_2 - x}{x_2 - x_1}\right)u_1 + \left(\frac{x - x_1}{x_2 - x_1}\right)u_2, \text{ for } x_1 \le x \le x_2 \\ &= N_1^{(1)}u_1 + N_2^{(1)}u_2\end{aligned} \tag{7.3}$$

where

$$\begin{aligned}N_1^{(1)} &= \frac{x_2 - x}{x_2 - x_1} \\ N_2^{(1)} &= \frac{x - x_1}{x_2 - x_1}\end{aligned} \tag{7.4}$$

The functions $N_1^{(1)}$ and $N_2^{(1)}$ are called the *shape functions for element one*. They define the type of interpolation, between nodal values, used within the element. As we discussed earlier, it is possible to use higher order interpolation such as quadratic or cubic. We will consider only linear shape functions, but the principles developed here remain valid for higher order interpolation.

Equation (7.3) may be generalized as follows. For element k, defined by nodes i and j which have x-coordinates x_i and x_j respectively, we have:

$$\begin{aligned}\tilde{u}^{(k)}(x) &= \left(\frac{x_j - x}{x_j - x_i}\right)u_i + \left(\frac{x - x_i}{x_j - x_i}\right)u_j, \quad \text{for } x_i \le x \le x_j \\ &= N_i^{(k)}u_i + N_j^{(k)}u_j\end{aligned} \tag{7.5}$$

So far then we have written the approximate solution in terms of the nodal values u_i and u_j. In general these are not known and must be found. In the next section we describe a method for determining the "best" values of the nodal variables.

7.4 Finding Nodal Values

When discussing the approximation of the curve shown in Figure (7.19), we noted that the approximation depends on both the number and shape of the line segments and the accuracy with which the nodal values are determined. In this section we discuss the latter.

7.4.1 Weighted Residual Methods

Suppose we have a differential operator, $\mathcal{L}$. By this we mean that $\mathcal{L}(u)$ represents an expression involving the function u and its derivatives. For example, we may have

$$\mathcal{L}(u) = \frac{\partial^2 u}{\partial x^2} - 2u^2\frac{\partial u}{\partial x} + 5 \tag{7.6}$$

A general differential equation can then be defined as

$$\mathcal{L}(u) - f = 0 \tag{7.7}$$

where f is some known function of the independent variables.

Now suppose that we substitute some function $\tilde{u}$ which is not an exact solution to Eq. (7.7). Then the RHS of Eq. (7.7) will be non-zero. We denote this non-zero value by $R(\tilde{u})$. That is,

$$\mathcal{L}(\tilde{u}) - f = \mathcal{R}(\tilde{u}) \tag{7.8}$$

Weighted residual methods attempt to reduce $R(\tilde{u})$ to zero. If the residual is close to zero, then we should have a reasonable approximation to the exact solution.

7.4.2 Galerkin's Method

The Galerkin method, (sometimes known as the Bubnov-Galerkin method), is one form of weighted residual method. There are other approaches [BD] such as:

- the collocation method
- the sub-domain method
- the least squares method.

In the Galerkin method, we enforce the following requirement:

$$\int_D R(\tilde{u}) N_i^{(k)} dD = 0 \tag{7.9}$$

Here the $N_i^{(k)}$ are the element shape functions defined in Section 7.3 and D is the domain of the element. At this stage we will say no more about this method. We will use the Galerkin method to derive element equations later in Section 7.6.

7.5 Constraint Equations

This section describes the way constraints are used to alter systems of equations. Constraints arise in the finite element method during assembly of boundary conditions and in application of boundary conditions.

We introduce the idea of constraint equations by example on a system of three equations.

Suppose we need to solve the following three equations:

$$\begin{aligned} k_{11}u_1 + k_{12}u_2 + k_{13}u_3 &= F_1 \\ k_{21}u_1 + k_{22}u_2 + k_{23}u_3 &= F_2 \\ k_{31}u_1 + k_{32}u_2 + k_{33}u_3 &= F_3 \end{aligned} \tag{7.10}$$

for the u_i where the k_{ij} and F_i are known.

These equations may be written in matrix notation as follows:

$$\begin{bmatrix} k_{11} & k_{12} & k_{13} \\ k_{21} & k_{22} & k_{23} \\ k_{31} & k_{32} & k_{33} \end{bmatrix} \begin{Bmatrix} u_1 \\ u_2 \\ u_3 \end{Bmatrix} = \begin{Bmatrix} F_1 \\ F_2 \\ F_3 \end{Bmatrix} \tag{7.11}$$

We will refer to the square matrix with terms k_{ij} as the *stiffness matrix.*

A constraint equation for this set of equations describes a relationship between the unknowns u_1, u_2 and u_3. A general constraint equation is:

$$c_1 u_1 + c_2 u_2 + c_3 u_3 = b \tag{7.12}$$

where, c_1, c_2, c_3 and b are known constants.

If a constraint equation exists, u_1, u_2 and u_3 are no longer independent and any one may be written in terms of the other two. This simplifies the solution process considerably. For example, we will write u_1 in terms of u_2 and u_3. In principle it makes no difference which one we use, however, for numerical stability when using a computer to solve the equations, the u_i with the largest coefficient is often the best choice. Hence,

$$u_1 = \frac{b}{c_1} - \frac{c_2}{c_1} u_2 - \frac{c_3}{c_1} u_3 \tag{7.13}$$

We now substitute Eq. (7.13) in Eq. (7.10) and so eliminate u_1. That is,

$$\begin{aligned} k_{11}\left(\frac{b}{c_1} - \frac{c_2}{c_1}u_2 - \frac{c_3}{c_1}u_3\right) + k_{12}u_2 + k_{13}u_3 &= F_1 \\ k_{21}\left(\frac{b}{c_1} - \frac{c_2}{c_1}u_2 - \frac{c_3}{c_1}u_3\right) + k_{22}u_2 + k_{23}u_3 &= F_2 \\ k_{31}\left(\frac{b}{c_1} - \frac{c_2}{c_1}u_2 - \frac{c_3}{c_1}u_3\right) + k_{32}u_2 + k_{33}u_3 &= F_3 \end{aligned} \tag{7.14}$$

Rearranging we get:

$$\begin{aligned}
\left(k_{12}-\frac{c_2}{c_1}k_{11}\right)u_2+\left(k_{13}-\frac{c_3}{c_1}k_{11}\right)u_3&=F_1-\frac{b}{c_1}k_{11}\\
\left(k_{22}-\frac{c_2}{c_1}k_{21}\right)u_2+\left(k_{23}-\frac{c_3}{c_1}k_{21}\right)u_3&=F_2-\frac{b}{c_1}k_{21}\\
\left(k_{32}-\frac{c_2}{c_1}k_{31}\right)u_2+\left(k_{33}-\frac{c_3}{c_1}k_{31}\right)u_3&=F_3-\frac{b}{c_1}k_{31}
\end{aligned}\qquad\textbf{(7.15)}$$

This may be written in matrix form as:

$$\begin{bmatrix} k_{12}-(c_2/c_1)k_{11} & k_{13}-(c_3/c_1)k_{11}\\ k_{22}-(c_2/c_1)k_{21} & k_{23}-(c_3/c_1)k_{21}\\ k_{32}-(c_2/c_1)k_{31} & k_{33}-(c_3/c_1)k_{31}\end{bmatrix}\begin{Bmatrix}u_2\\u_3\end{Bmatrix}=\begin{Bmatrix}F_1-(b/c_1)k_{11}\\F_2-(b/c_1)k_{21}\\F_3-(b/c_1)k_{31}\end{Bmatrix}\qquad\textbf{(7.16)}$$

Comparing Eq. (7.16) with Eq. (7.11), the procedure used to arrive at Eq. (7.16) can be considered as a set of column operations on Eq. (7.11). The operations are:

(7.17)

1. Multiply the first column of the stiffness matrix by $-c_2/c_1$ and add to the second column
2. Multiply the first column of the stiffness matrix by $-c_3/c_1$ and add to the third column.
3. Multiply the first column of the stiffness matrix by b/c_1 and subtract from the RHS matrix.
4. Delete the first column of the stiffness matrix and u_1.

A similar set of operations can then be performed on the rows of Eq. (7.16):

(7.18)

1. Multiply the first column of the new stiffness matrix and the RHS matrix by $-c_2/c_1$ and add to the second row.
2. Multiply the first column of the new stiffness matrix and the RHS matrix by $-c_3/c_1$ and add to the third row.
3. Delete the first row of the stiffness matrix and the first row of the RHS matrix.

Applying the row operations (7.18) to Eq. (7.16) gives:

$$\begin{bmatrix} k_{22} - \frac{c_2}{c_1}k_{21} - \frac{c_2}{c_1}\left(k_{12} - \frac{c_2}{c_1}k_{11}\right) & k_{23} - \frac{c_3}{c_1}k_{21} - \frac{c_2}{c_1}\left(k_{13} - \frac{c_3}{c_1}k_{11}\right) \\ k_{32} - \frac{c_2}{c_1}k_{31} - \frac{c_3}{c_1}\left(k_{12} - \frac{c_2}{c_1}k_{11}\right) & k_{33} - \frac{c_3}{c_1}k_{31} - \frac{c_3}{c_1}\left(k_{13} - \frac{c_3}{c_1}k_{11}\right) \end{bmatrix} \begin{Bmatrix} u_2 \\ u_3 \end{Bmatrix} = \begin{Bmatrix} \left(F_2 - \frac{b}{c_1}k_{21}\right) - \frac{c_2}{c_1}\left(F_1 - \frac{b}{c_1}k_{11}\right) \\ \left(F_3 - \frac{b}{c_1}k_{31}\right) - \frac{c_3}{c_1}\left(F_1 - \frac{b}{c_1}k_{11}\right) \end{Bmatrix} \quad \textbf{(7.19)}$$

The original problem has now been reduced to two equations in two unknowns. This may be solved for u_2 and u_3. Finally, u_1 is determined by substitution in the constraint equation Eq. (7.13). Note that although this type of manipulation is tedious when done manually, it is very easy to implement in a computer program

To conclude this section we consider two special cases of constraint equations. They are:

1. when two of the constraints are equal. That is the constraint equation takes the form; $u_i = u_j,\ i \neq j$, and
2. when one of the unknowns is constant. That is the constraint equation has the form; $u_i = c$.

We will use both these types of constraint equation in later sections when discussing assembly of equations and setting boundary conditions.

7.5.1 Special Case 1: Two Unknowns Equal

In this case we have the constraint equation $u_i = u_j$. This type of constraint equation is important when assembling the element equations to find the system equations.

To make the example concrete we will consider the equations:

$$\begin{bmatrix} k_{11}^{(1)} & k_{12}^{(1)} & 0 & 0 \\ k_{21}^{(1)} & k_{22}^{(1)} & 0 & 0 \\ 0 & 0 & k_{11}^{(2)} & k_{12}^{(2)} \\ 0 & 0 & k_{21}^{(2)} & k_{22}^{(2)} \end{bmatrix} \begin{Bmatrix} u_1 \\ u_2 \\ u_3 \\ u_4 \end{Bmatrix} = \begin{Bmatrix} F_1 \\ F_2 \\ F_3 \\ F_4 \end{Bmatrix} \quad \textbf{(7.20)}$$

subject to the constraint equation

$$u_2 = u_3 \quad \textbf{(7.21)}$$

Although the notation for the entries in the matrix on the LHS of Eq. (7.20) may appear odd, it is will be explained later. For now just take the entries to be unique numbers.

Writing the constraint in the form of Eq. (7.13) we get

$$u_2 = 0 + 0.u_1 + 1.u_3 \tag{7.22}$$

The column operations (7.17) defined earlier become:

1. Multiply the second column of the stiffness matrix in Eq. (7.20) by 0 and add to the first column.
2. Multiply the second column of the stiffness matrix in Eq. (7.20) by 1 and add to the third column.
3. Multiply the second column of the stiffness matrix in Eq. (7.20) by 0 and add to the RHS.
4. Delete the second column of the stiffness matrix in Eq. (7.20) and u_2.

Of these, only list items 2 and 4 cause any change. Applying these to Eq. (7.20) we get:

$$\begin{bmatrix} k_{11}^{(1)} & k_{12}^{(1)} & 0 \\ k_{21}^{(1)} & k_{22}^{(1)} & 0 \\ 0 & k_{11}^{(2)} & k_{12}^{(2)} \\ 0 & k_{21}^{(2)} & k_{22}^{(2)} \end{bmatrix} \begin{Bmatrix} u_1 \\ u_3 \\ u_4 \end{Bmatrix} = \begin{Bmatrix} F_1 \\ F_2 \\ F_3 \\ F_4 \end{Bmatrix} \tag{7.23}$$

The row operations (7.18) defined earlier become:

1. Multiply the second row of the stiffness matrix and the RHS matrix in Eq. (7.23) by 0 and add to the first row.
2. Multiply the second row of the stifness matrix and the RHS matrix in Eq. (7.23) by 1 and add to the third row.
3. Delete the second row of stiffness matrix and RHS matrix.

Here, only items 2 and 3 have an effect. Applying these operations to Eq. (7.23) gives:

$$\begin{bmatrix} k_{11}^{(1)} & k_{21}^{(1)} & 0 \\ k_{21}^{(1)} & k_{22}^{(1)} + k_{11}^{(2)} & k_{12}^{(2)} \\ 0 & k_{21}^{(2)} & k_{22}^{(2)} \end{bmatrix} \begin{Bmatrix} u_1 \\ u_3 \\ u_4 \end{Bmatrix} = \begin{Bmatrix} F_1 \\ F_2 + F_3 \\ F_4 \end{Bmatrix} \tag{7.24}$$

This may then be solved for u_1, u_3 and u_4. The value of u_2 is then obtained from Eq. (7.21).

7.5.2 Special Case 2: One Unknown Constant

In this case the constraint equation is u_i = constant. This type of constraint equation arises when applying boundary conditions.

As an example we consider the equations:

$$\begin{bmatrix} k_{11} & k_{12} & k_{13} \\ k_{21} & k_{22} & k_{23} \\ k_{31} & k_{32} & k_{33} \end{bmatrix} \begin{Bmatrix} u_1 \\ u_2 \\ u_3 \end{Bmatrix} = \begin{Bmatrix} F_1 \\ F_2 \\ F_3 \end{Bmatrix} \tag{7.25}$$

with the constraint equation $u_2 = c$. Writing the constraint in the form of Eq. (7.13) we get:

$$u_2 = c + 0.u_1 + 0.u_3 \tag{7.26}$$

Steps 1 and 2 of the column operations (7.17) have no effect. Only operations 3 and 4 need be done. These become:

1. Multiply column 2 of the stiffness matrix by c and subtract from the RHS.
2. Delete second column of stiffness matrix and u_2.

Applying these to Eq. (7.25) gives:

$$\begin{bmatrix} k_{11} & k_{13} \\ k_{21} & k_{23} \\ k_{31} & k_{33} \end{bmatrix} \begin{Bmatrix} u_1 \\ u_3 \end{Bmatrix} = \begin{Bmatrix} F_1 - ck_{12} \\ F_2 - ck_{22} \\ F_3 - ck_{32} \end{Bmatrix} \tag{7.27}$$

Steps 1 and 2 of the row operations (7.18) have no effect. Only step 3 need be done. This becomes:

1. Delete second row of the stiffness matrix and the RHS matrix.

Hence Eq. (7.27) becomes:

$$\begin{bmatrix} k_{11} & k_{13} \\ k_{31} & k_{33} \end{bmatrix} \begin{Bmatrix} u_1 \\ u_3 \end{Bmatrix} = \begin{Bmatrix} F_1 - ck_{12} \\ F_3 - ck_{32} \end{Bmatrix} \tag{7.28}$$

This may be solved for u_1 and u_3.

Note that if $c = 0$, that is, $u_2 = 0$, then Eq. (7.28) becomes:

$$\begin{bmatrix} k_{11} & k_{13} \\ k_{31} & k_{33} \end{bmatrix} \begin{Bmatrix} u_1 \\ u_3 \end{Bmatrix} = \begin{Bmatrix} F_1 \\ F_3 \end{Bmatrix} \tag{7.29}$$

Comparing this with the original equation Eq. (7.25) we see that the constraint $u_2 = 0$ may be applied by deleting the second row and second column from each of the matrices in Eq. (7.25). In general, the constraint $u_i = 0$, is applied by deleting the ith row and columns from the relevant matrices.

7.6 A One Dimensional Problem Solved Using the FEM

In this section we illustrate many of the important concepts of finite element analysis in a simple example.

We will solve the following problem:

$$\frac{d^2u}{dx^2} - 2 = 0, \text{ for } 1 \leq x \leq 4 \tag{7.30}$$

subject to the boundary conditions:

$$\begin{aligned} \frac{du}{dx} &= 3, \text{ at } x = 1 \\ u(x) &= 20, \text{ at } x = 4 \end{aligned} \tag{7.31}$$

This equation may be solved exactly by integrating twice. The solution is

$$u(x) = x^2 + x \tag{7.32}$$

and is shown graphically in Fig. (7.21). After using the finite element method to solve the problem we will compare the approximate solution with the exact solution.

Figure 7.21: Exact Solution to Problem

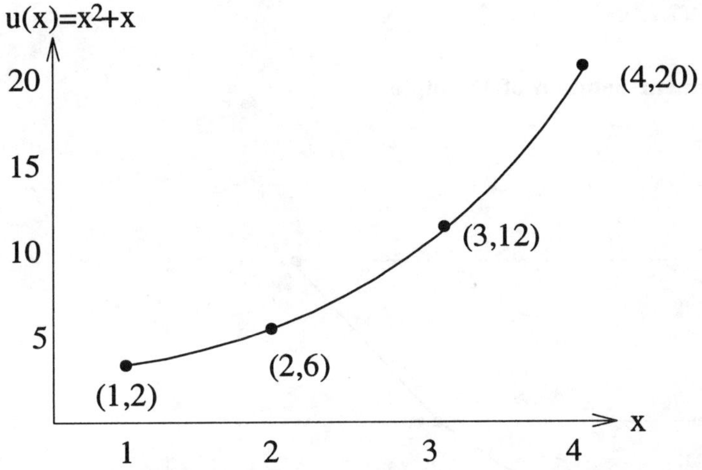

7.6.1 Meshing

The domain for the problem is $\{x : 1 \leq x \leq 4\}$. For finite element solution we will break this up into three elements, each of which is defined by two nodes, as shown in Fig. (7.22).

Figure 7.22: Mesh Definition for FEM Example

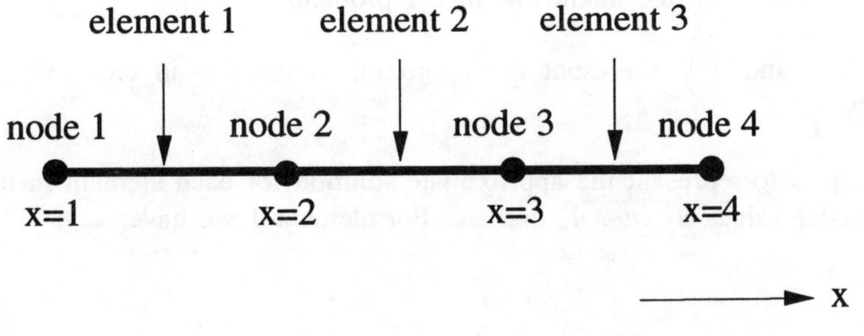

In total then, we have four nodes defining three elements. There will be four nodal values to determine.

7.6.2 Derivation of Element Equations

We will use linear shape functions for the example. That is, we will seek a solution of the type shown in Fig. (7.23).

Figure 7.23: Finite Element Solution of Example

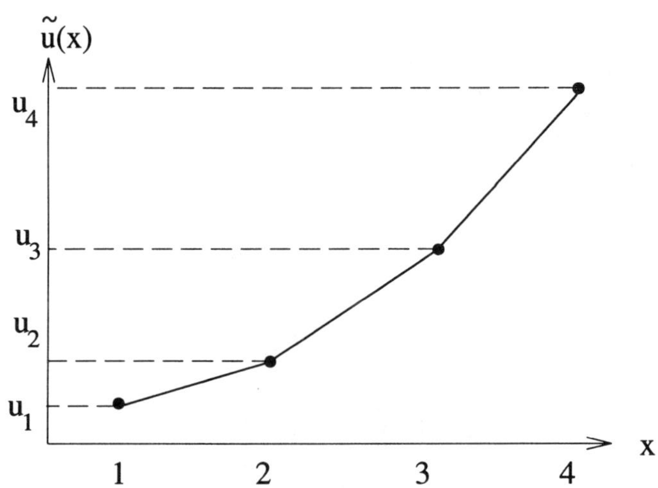

Let u_1, u_2, u_3 and u_4 be the values of the approximate solution at nodes 1, 2, 3 and 4 respectively. These are the unknowns in our problem.

Let $\tilde{u}^{(1)}$, $\tilde{u}^{(2)}$ and $\tilde{u}^{(3)}$ represent the approximate solution in elements 1, 2 and 3 respectively.

The first step is to represent the approximate solution for each element in terms of the unknown nodal values u_1, u_2, u_3 and u_4. For element 1 we have:

$$\begin{aligned}\tilde{u}^{(1)}(x) &= (2-x)u_1 + (x-1)u_2 \\ &= N_1^{(1)}u_1 + N_2^{(1)}u_2\end{aligned} \qquad \textbf{(7.33)}$$

where

$$\begin{aligned} N_1^{(1)} &= 2 - x \\ N_2^{(1)} &= x - 1 \end{aligned} \qquad \textbf{(7.34)}$$

The functions $N_1^{(1)}$ and $N_2^{(1)}$ are the shape functions for element 1 and were defined in Section 7.3.

Similarly for element 2 we write:

$$\begin{aligned} \tilde{u}^{(2)}(x) &= (3 - x)u_2 + (x - 2)u_3 \\ &= N_1^{(2)}u_2 + N_2^{(2)}u_3 \end{aligned} \qquad \textbf{(7.35)}$$

where

$$\begin{aligned} N_1^{(2)} &= 3 - x \\ N_2^{(2)} &= x - 2 \end{aligned} \qquad \textbf{(7.36)}$$

And for element 3 we have,

$$\begin{aligned} \tilde{u}^{(3)}(x) &= (4 - x)u_3 + (x - 3)u_4 \\ &= N_1^{(3)}u_3 + N_2^{(3)}u_4 \end{aligned} \qquad \textbf{(7.37)}$$

where

$$\begin{aligned} N_1^{(3)} &= 4 - x \\ N_2^{(3)} &= x - 3 \end{aligned} \qquad \textbf{(7.38)}$$

The residual for each element is given by:

$$\begin{aligned} R\Big(\tilde{u}^{(e)}(x)\Big) &= \frac{d^2}{dx^2}\Big(\tilde{u}^{(e)}\Big) - 2 \\ &= \frac{d}{dx}\left(\frac{d\tilde{u}^{(e)}}{dx}\right) - 2 \quad \text{where } e = 1, 2, 3. \end{aligned} \qquad \textbf{(7.39)}$$

Galerkin's method (see Section 7.4.2) requires that for each element,

$$\int_{x_a}^{x_b} R\Big(\tilde{u}^{(e)}(x)\Big)N_i^{(e)}dx = 0 \quad i = 1,2 \tag{7.40}$$

where x_a and x_b are the x-coordinates for the nodes defining element number e. Note that for each element, Eq. (7.40) gives two equations.

Substituting Eq. (7.39) in Eq. (7.40), we get

$$\begin{aligned} 0 &= \int_{x_a}^{x_b} \left[\frac{d}{dx}\left(\frac{d\tilde{u}^{(e)}}{dx}\right) - 2\right] N_i^{(e)} dx \\ &= \int_{x_a}^{x_b} \frac{d}{dx}\left(\frac{d\tilde{u}^{(e)}}{dx}\right) N_i^{(e)} dx - \int_{x_a}^{x_b} 2N_i^{(e)} dx \end{aligned} \tag{7.41}$$

The first term on the RHS may be integrated by parts to give:

$$\int_{x_a}^{x_b} \frac{d}{dx}\left(\frac{d\tilde{u}^{(e)}}{dx}\right) N_i^{(e)} dx = \left[\frac{d\tilde{u}^{(e)}}{dx} N_i^{(e)}\right]_{x=x_a}^{x=x_b} - \int_{x_a}^{x_b} \frac{d\tilde{u}^{(e)}}{dx}\frac{dN_i^{(e)}}{dx} dx \tag{7.42}$$

Substituting Eq. (7.42) into Eq. (7.41) then gives:

$$0 = \left[\frac{d\tilde{u}^{(e)}}{dx} N_i^{(e)}\right]_{x=x_a}^{x=x_b} - \int_{x_a}^{x_b} \frac{d\tilde{u}^{(e)}}{dx}\frac{dN_i^{(e)}}{dx} dx - \int_{x_a}^{x_b} 2N_i^{(e)} dx \tag{7.43}$$

Rearranging we have

$$\int_{x_a}^{x_b} \frac{d\tilde{u}^{(e)}}{dx}\frac{dN_i^{(e)}}{dx} dx = \left[\frac{d\tilde{u}^{(e)}}{dx} N_i^{(e)}\right]_{x=x_a}^{x=x_b} - \int_{x_a}^{x_b} 2N_i^{(e)} dx \tag{7.44}$$

Substituting $\tilde{u}^{(1)}$ from Eq. (7.33) into the LHS of Eq. (7.44) gives:

$$\begin{aligned}\int_1^2 \frac{d\tilde{u}^{(1)}}{dx}\frac{dN_i^{(1)}}{dx}dx &= \int_1^2 \frac{d}{dx}\left(N_1^{(1)}u_1 + N_2^{(1)}u_2\right)\frac{dN_i^{(1)}}{dx}dx \\ &= \int_1^2 \frac{dN_1^{(1)}}{dx}u_1\frac{dN_i^{(1)}}{dx}dx + \int_1^2 \frac{dN_2^{(1)}}{dx}u_2\frac{dN_i^{(1)}}{dx}dx \\ &= \left(\int_1^2 \frac{dN_1^{(1)}}{dx}\frac{dN_i^{(1)}}{dx}dx\right)u_1 + \left(\int_1^2 \frac{dN_2^{(1)}}{dx}\frac{dN_i^{(1)}}{dx}dx\right)u_2\end{aligned} \tag{7.45}$$

Substituting the above in Eq. (7.44) we find that, for element one, Eq. (7.44) gives the following two equations corresponding to $i = 1$ and $i = 2$ respectively,

$$\begin{aligned}\left(\int_1^2 \frac{dN_1^{(1)}}{dx}\frac{dN_1^{(1)}}{dx}dx\right)u_1 + \left(\int_1^2 \frac{dN_2^{(1)}}{dx}\frac{dN_1^{(1)}}{dx}dx\right)u_2 &= \left[\frac{d\tilde{u}^{(1)}}{dx}N_1^{(1)}\right]_{x=1}^{x=2} \\ &\quad - \int_1^2 2N_1^{(1)}dx \\ \left(\int_1^2 \frac{dN_1^{(1)}}{dx}\frac{dN_2^{(1)}}{dx}dx\right)u_1 + \left(\int_1^2 \frac{dN_2^{(1)}}{dx}\frac{dN_2^{(1)}}{dx}dx\right)u_2 &= \left[\frac{d\tilde{u}^{(1)}}{dx}N_2^{(1)}\right]_{x=1}^{x=2} \\ &\quad - \int_1^2 2N_2^{(1)}dx\end{aligned} \tag{7.46}$$

Equation (7.46) can be written in matrix form as follows:

$$\begin{bmatrix} k_{11}^{(1)} & k_{12}^{(1)} \\ k_{21}^{(1)} & k_{22}^{(1)} \end{bmatrix}\begin{Bmatrix} u_1 \\ u_2 \end{Bmatrix} = \begin{Bmatrix} F_1^{(1)} \\ F_2^{(1)} \end{Bmatrix} \tag{7.47}$$

where

$$k_{11}^{(1)} = \int_1^2 \frac{dN_1^{(1)}}{dx}\frac{dN_1^{(1)}}{dx}dx = 1$$

$$k_{12}^{(1)} = \int_1^2 \frac{dN_2^{(1)}}{dx}\frac{dN_1^{(1)}}{dx}dx = -1$$

$$k_{21}^{(1)} = \int_1^2 \frac{dN_1^{(1)}}{dx}\frac{dN_2^{(1)}}{dx}dx = -1$$

$$k_{22}^{(1)} = \int_1^2 \frac{dN_2^{(1)}}{dx}\frac{dN_2^{(1)}}{dx}dx = 1 \qquad \textbf{(7.48)}$$

$$F_1^{(1)} = \left[\frac{d\tilde{u}^{(1)}}{dx}N_1^{(1)}\right]_{x=1}^{x=2} - \int_1^2 2N_1^{(1)}dx$$

$$= -\left(\frac{d\tilde{u}^{(1)}}{dx}\right)_{x=1} - 1$$

$$F_2^{(1)} = \left[\frac{d\tilde{u}^{(1)}}{dx}N_2^{(1)}\right]_{x=1}^{x=2} - \int_1^2 2N_2^{(1)}dx$$

$$= \left(\frac{d\tilde{u}^{(1)}}{dx}\right)_{x=2} - 1$$

In the above, the terms $k_{ij}^{(1)}$ and $F_i^{(1)}$ were evaluated by first substituting for $\tilde{u}^{(1)}$ and $N_i^{(1)}$ using equations (7.33) and (7.34) respectively, and then carrying out the integrations.

Equations (7.47) are the element equations for element one. The square matrix on the LHS of (7.47) is called the stiffness matrix for element one. An important property of the matrix is its symmetry, that is $k_{ij} = k_{ji}$.

Using precisely the same steps used to produce Eq. (7.47), it is possible to write similar equations for elements two and three. In fact the only changes required for the other elements are

i. Replace the superscript $^{(1)}$ by $^{(2)}$ or $^{(3)}$,

ii. Change the limits of the integration, and

iii. Change the subscripts on the nodal unknowns.

For element two we get:

$$\begin{bmatrix} k_{11}^{(2)} & k_{12}^{(2)} \\ k_{21}^{(2)} & k_{22}^{(2)} \end{bmatrix} \begin{Bmatrix} u_2 \\ u_3 \end{Bmatrix} = \begin{Bmatrix} F_1^{(2)} \\ F_2^{(2)} \end{Bmatrix} \tag{7.49}$$

where

$$\begin{aligned}
k_{11}^{(2)} &= \int_2^3 \frac{dN_1^{(2)}}{dx} \frac{dN_1^{(2)}}{dx} dx = 1 \\
k_{12}^{(2)} &= \int_2^3 \frac{dN_2^{(2)}}{dx} \frac{dN_1^{(2)}}{dx} dx = -1 \\
& k_{21}^{(2)} \\
k_{22}^{(2)} &= \int_2^3 \frac{dN_2^{(2)}}{dx} \frac{dN_2^{(2)}}{dx} dx = 1 \\
F_1^{(2)} &= \left[\frac{d\tilde{u}^{(2)}}{dx} N_1^{(2)} \right]_{x=2}^{x=3} - \int_2^3 2N_1^{(2)} dx \\
&= -\left(\frac{d\tilde{u}^{(2)}}{dx} \right)_{x=2} - 1 \\
F_2^{(2)} &= \left[\frac{d\tilde{u}^{(2)}}{dx} N_2^{(2)} \right]_{x=2}^{x=3} - \int_2^3 2N_2^{(2)} dx \\
&= \left(\frac{d\tilde{u}^{(2)}}{dx} \right)_{x=3} - 1
\end{aligned} \tag{7.50}$$

Finally, for element three we get:

$$\begin{bmatrix} k_{11}^{(3)} & k_{12}^{(3)} \\ k_{21}^{(3)} & k_{22}^{(3)} \end{bmatrix} \begin{Bmatrix} u_3 \\ u_4 \end{Bmatrix} = \begin{Bmatrix} F_1^{(3)} \\ F_2^{(3)} \end{Bmatrix} \tag{7.51}$$

where

$$\begin{aligned}
k_{11}^{(3)} &= \int_3^4 \frac{dN_1^{(3)}}{dx}\frac{dN_1^{(3)}}{dx}dx = 1 \\
k_{12}^{(3)} &= \int_3^4 \frac{dN_2^{(3)}}{dx}\frac{dN_1^{(3)}}{dx}dx = -1 \\
k_{21}^{(3)} &= \int_3^4 \frac{dN_2^{(3)}}{dx}\frac{dN_1^{(3)}}{dx}dx = -1 \\
k_{22}^{(3)} &= \int_3^4 \frac{dN_2^{(3)}}{dx}\frac{dN_2^{(3)}}{dx}dx = 1 \\
F_1^{(3)} &= \left[\frac{d\tilde{u}^{(3)}}{dx}N_1^{(3)}\right]_{x=3}^{x=4} - \int_3^4 2N_1^{(3)}dx \\
&= -\left(\frac{d\tilde{u}^{(3)}}{dx}\right)_{x=3} - 1 \\
F_2^{(3)} &= \left[\frac{d\tilde{u}^{(3)}}{dx}N_2^{(3)}\right]_{x=3}^{x=4} - \int_3^4 2N_2^{(3)}dx \\
&= \left(\frac{d\tilde{u}^{(3)}}{dx}\right)_{x=4} - 1
\end{aligned} \tag{7.52}$$

This completes the derivation of the element equations. Though tedious by hand, the procedure is easily implemented on a computer. Before considering the assembly of these equations it is worth revising the main points of this section.

First we chose linear interpolation for the value of $\tilde{u}$ between two nodes. This defined the shape functions for each element. The Galerkin method was then applied to each element. This resulted in two algebraic equations for each element; equations (7.47), (7.49) and (7.51). This last point is crucial. The Galerkin method produced algebraic equations for the unknown u_i. In effect then, this technique reduces a differential equation to a number of algebraic equations.

7.6.3 Assembly

For the assembly process we will superscript the nodal unknowns to indicate which element they are associated with. That is, $u_1^{(1)}$ indicates the value of the approximate solution at node one which is associated with element one, $u_2^{(1)}$ indicates the value of the approximate solution at node two which is associated with element one, and so on.

The element equations (7.47), (7.49) and (7.51) can be written as a single matrix equation as shown below:

$$\begin{bmatrix} k_{11}^{(1)} & k_{12}^{(1)} & 0 & 0 & 0 & 0 \\ k_{21}^{(1)} & k_{22}^{(1)} & 0 & 0 & 0 & 0 \\ 0 & 0 & k_{11}^{(2)} & k_{12}^{(2)} & 0 & 0 \\ 0 & 0 & k_{21}^{(2)} & k_{22}^{(2)} & 0 & 0 \\ 0 & 0 & 0 & 0 & k_{11}^{(3)} & k_{12}^{(3)} \\ 0 & 0 & 0 & 0 & k_{21}^{(3)} & k_{22}^{(3)} \end{bmatrix} \begin{Bmatrix} u_1^{(1)} \\ u_2^{(1)} \\ u_2^{(2)} \\ u_3^{(2)} \\ u_3^{(3)} \\ u_4^{(3)} \end{Bmatrix} = \begin{Bmatrix} F_1^{(1)} \\ F_2^{(1)} \\ F_1^{(2)} \\ F_2^{(2)} \\ F_1^{(3)} \\ F_2^{(3)} \end{Bmatrix} \tag{7.53}$$

Now consider the terms $u_2^{(1)}$ and $u_2^{(2)}$. These are the values of the approximate solution at $x = 2$ for elements one and two respectively. Substituting $x = 2$ in equations (7.33) and (7.35) gives:

$$\begin{aligned} \tilde{u}_1^{(1)} &= \tilde{u}^{(1)}(2) \\ &= (2-1)u_2 \\ &= u_2 \\ \tilde{u}_2^{(2)} &= \tilde{u}^{(2)}(2) \\ &= (3-2)u_2 \\ &= u_2 \end{aligned} \tag{7.54}$$

Therefore,

$$u_2^{(1)} = u_2^{(2)} = u_2 \tag{7.55}$$

That this is so comes as no great surprise. Intuitively, any solution to the problem should be continuous across the node at $x = 2$. This requirement is called an *inter-element boundary condition (IBC)*. Assembly of the element equations is simply the process of enforcing the IBC's.

In a similar way we find from equations (7.35) and (7.37) that

$$u_3^{(2)} = u_3^{(3)} = u_3 \tag{7.56}$$

Now equations (7.55) and (7.56) are constraint equations on the system of equations (7.53). Applying the column and row operations given in Section 7.5.1 to Eq. (7.53), we enforce the IBC's and so assemble the element equations. We obtain

$$\begin{bmatrix} k_{11}^{(1)} & k_{12}^{(1)} & 0 & 0 \\ k_{21}^{(1)} & k_{22}^{(1)} + k_{11}^{(2)} & k_{12}^{(2)} & 0 \\ 0 & k_{21}^{(2)} & k_{22}^{(2)} + k_{11}^{(3)} & k_{12}^{(3)} \\ 0 & 0 & k_{21}^{(3)} & k_{22}^{(3)} \end{bmatrix} \begin{Bmatrix} u_1 \\ u_2 \\ u_3 \\ u_4 \end{Bmatrix} = \begin{Bmatrix} F_1^{(1)} \\ F_2^{(1)} + F_1^{(2)} \\ F_2^{(2)} + F_1^{(3)} \\ F_2^{(3)} \end{Bmatrix} \tag{7.57}$$

where we have now dropped the superscripts on the unknowns.

These equations are the system equations. They represent the FEM solution to our problem over the whole domain.

Before applying boundary conditions we substitute the values for the terms in equation (7.57). These were given in equations (7.48) to (7.52). Equation (7.57) therefore becomes:

$$\begin{bmatrix} 1 & -1 & 0 & 0 \\ -1 & 2 & -1 & 0 \\ 0 & -1 & 2 & -1 \\ 0 & 0 & -1 & 0 \end{bmatrix} \begin{Bmatrix} u_1 \\ u_2 \\ u_3 \\ u_4 \end{Bmatrix} = \begin{Bmatrix} -\left(d\tilde{u}^{(1)}/dx\right)_{x=1} - 1 \\ \left(d\tilde{u}^{(1)}/dx\right)_{x=2} - \left(d\tilde{u}^{(2)}/dx\right)_{x=2} - 2 \\ \left(d\tilde{u}^{(2)}/dx\right)_{x=3} - \left(d\tilde{u}^{(3)}/dx\right)_{x=3} - 2 \\ \left(d\tilde{u}^{(3)}/dx\right)_{x=4} - 1 \end{Bmatrix} \tag{7.58}$$

7.6.4 Application of Boundary Conditions

The boundary conditions were given in equation (7.31). They are:

$$\begin{aligned} \frac{du}{dx} &= 3, \text{ at } x = 1 \\ u(x) &= 20, \text{ at } x = 4 \end{aligned} \tag{7.59}$$

The first of these is applied by direct substitution in Eq. (7.58). We obtain

$$\begin{bmatrix} 1 & -1 & 0 & 0 \\ -1 & 2 & -1 & 0 \\ 0 & -1 & 2 & -1 \\ 0 & 0 & -1 & 0 \end{bmatrix} \begin{Bmatrix} u_1 \\ u_2 \\ u_3 \\ u_4 \end{Bmatrix} = \begin{Bmatrix} -4 \\ \left(d\tilde{u}^{(1)}/dx\right)_{x=2} - \left(d\tilde{u}^{(2)}/dx\right)_{x=2} - 2 \\ \left(d\tilde{u}^{(2)}/dx\right)_{x=3} - \left(d\tilde{u}^{(3)}/dx\right)_{x=3} - 2 \\ \left(d\tilde{u}^{(3)}/dx\right)_{x=4} - 1 \end{Bmatrix} \tag{7.60}$$

The next boundary condition, $u(4) = 20$, is a constraint equation and is applied using the row and column operations of Section 7.5.2. That is, we multiply column four of the stiffness matrix in Eq. (7.60) by 20, subtract this from the RHS matrix, delete column four of the stiffness matrix, delete u_4 and finally, delete row four of the stiffness matrix and the RHS matrix. This gives:

$$\begin{bmatrix} 1 & -1 & 0 \\ -1 & 2 & -1 \\ 0 & -1 & 2 \end{bmatrix} \begin{Bmatrix} u_1 \\ u_2 \\ u_3 \end{Bmatrix} = \begin{Bmatrix} -4 \\ \left(d\tilde{u}^{(1)}/dx\right)_{x=2} - \left(d\tilde{u}^{(2)}/dx\right)_{x=2} - 2 \\ \left(d\tilde{u}^{(2)}/dx\right)_{x=3} - \left(d\tilde{u}^{(3)}/dx\right)_{x=3} + 18 \end{Bmatrix} \tag{7.61}$$

We now consider the RHS terms:

$$\begin{aligned} &\left(d\tilde{u}^{(1)}/dx\right)_{x=2} - \left(d\tilde{u}^{(2)}/dx\right)_{x=2} \\ &\left(d\tilde{u}^{(2)}/dx\right)_{x=3} - \left(d\tilde{u}^{(3)}/dx\right)_{x=3} \end{aligned} \tag{7.62}$$

Since our problem involves a second derivative, the first derivative, du/dx must be continuous over the solution domain. In our example the first derivatives of the approximate solution $d\tilde{u}/dx$ will not be continuous across the nodes due to our use of linear shape functions. Despite this, we enforce the continuity of first derivatives on our approximate solution. That is, we set

$$\begin{aligned} \left(\frac{d\tilde{u}^{(1)}}{dx}\right)_{x=2} &= \left(\frac{d\tilde{u}^{(2)}}{dx}\right)_{x=2} \\ \left(\frac{d\tilde{u}^{(2)}}{dx}\right)_{x=3} &= \left(\frac{d\tilde{u}^{(3)}}{dx}\right)_{x=3} \end{aligned} \tag{7.63}$$

Equations (7.63) are referred to as *natural boundary conditions*. They arise "naturally" in the finite element formulation. Substitution of Eq. (7.63) into the RHS of (7.61) gives:

$$\begin{bmatrix} 1 & -1 & 0 \\ -1 & 2 & -1 \\ 0 & -1 & 2 \end{bmatrix} \begin{Bmatrix} u_1 \\ u_2 \\ u_3 \end{Bmatrix} = \begin{Bmatrix} -4 \\ -2 \\ 18 \end{Bmatrix} \tag{7.64}$$

These equations are the final equations for the finite element method solution. All we need do now is solve them.

7.6.5 Solution of System Equations

Equations (7.64) can be solved by elementary methods as follows. First write the equations separately as:

$$\begin{aligned} u_1 - u_2 &= -4 \\ -u_1 + 2u_2 - u_3 &= -2 \\ -u_2 + 2u_3 &= 18 \end{aligned} \tag{7.65}$$

Adding these equations, we get, $u_3 = 12$. Substituting this in the third equation in (7.65) gives, $u_2 = 6$. Finally, substitution of this in the first equation gives $u_1 = 2$. From the boundary condition given in Eq. (7.31) we know $u_4 = 20$.

It should be noted that finite element problems frequently involve the solution of thousands of linear equations. For such cases the elementary method used above is not appropriate and a numerical procedure must be used. This is discussed later in Section 9.2.5.

Our approximate solution is determined by substituting the nodal values into equations (7.33), (7.35) and (7.37). We obtain:

$$\begin{aligned} \tilde{u}^{(1)}(x) &= (2-x).2 + (x-1).6 \\ &= 4x - 2 \quad \text{for} \quad 1 \le x \le 2 \\ \tilde{u}^{(2)}(x) &= (3-x).6 + (x-2).12 \\ &= 6x - 6 \quad \text{for} \quad 2 \le x \le 3 \\ \tilde{u}^{(3)}(x) &= (4-x).12 + (x-3).20 \\ &= 8x - 12 \quad \text{for} \quad 3 \le x \le 4 \end{aligned} \tag{7.66}$$

Both the exact solution and the approximate solution given in Eq. (7.66) are plotted in Figure (7.24).Note that in this range the exact and approximate solutions are identical at x=1, 2, 3, 4.

Figure 7.24: Comparison of Exact and FEM Solution

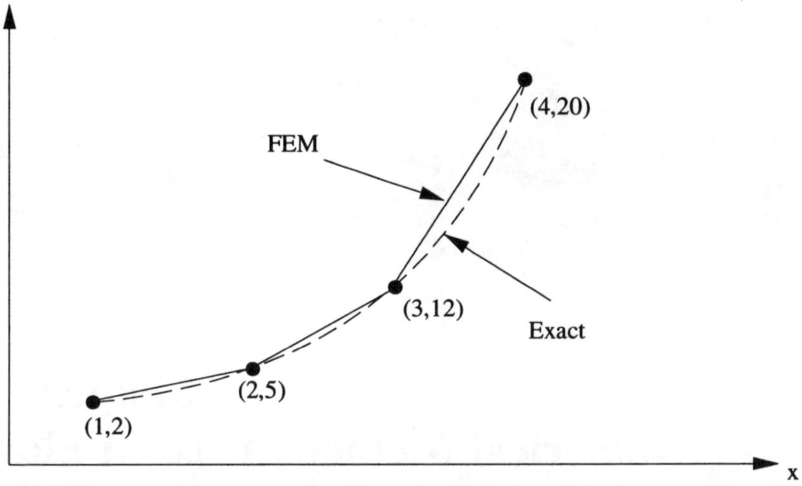

This concludes our introductory example. Later, in Chapters 9 and 10, we use the techniques of this Chapter to derive one and two dimensional elements for our mathematical model of injection molding.

7.7 Conclusion

It should be pointed out that use of finite element methods on the simple example above is a little like driving in a thumbtack with a sledge hammer. As you can see, the work involved in carrying out the finite element solution cannot be justified for simple problems. However, when there is no exact solution available for a problem the FEM comes into its own. While the hand calculations above are tedious, they are readily implemented on computers. It is this capability which makes the finite element method so attractive for large scale engineering problems.

Chapter 8
Numerical Methods Used in Flow Analysis: The Finite Difference Method

The finite difference method is one of the oldest numerical methods for producing approximate solutions to physical problems. In this Chapter we briefly review some fundamental concepts associated with finite difference methods. As in the previous Chapter, we don't attempt a complete presentation. We describe only what is necessary for an understanding of later Chapters.

There are many books on finite difference methods (see for example [SG], [AW], [FC]). An excellent introduction is given in [GS].

8.1 Introduction to the Finite Difference Method

Essentially the finite difference method discretizes a differential equation into a system of algebraic linear equations. The simplest approach to this process is via Taylor series expansion.

Just as in the finite element method, finite difference methods require a mesh. The mesh is generally some collection of points, called nodes, that covers the domain of interest. Figure (8.25) shows a section of mesh in the *x*-*y* plane.

Figure 8.25: Finite Difference Mesh

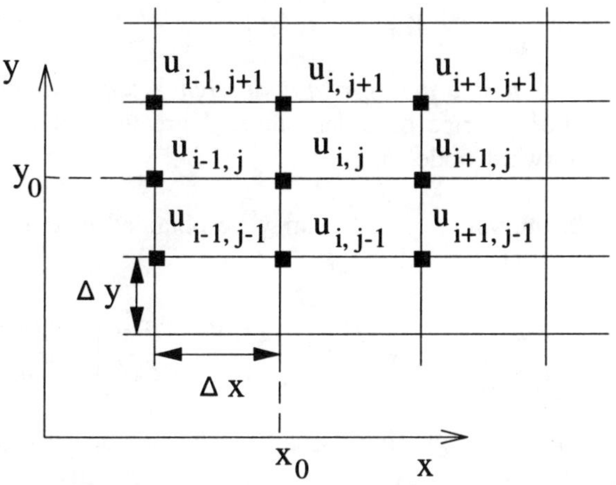

Suppose $u(x,y)$ is some function of x and y. We denote the value of $u(x,y)$ at the node (x_i, y_j) by $u_{i,j}$. If $u(x,y)$ is continuous, its value, $u_{i+1,j}$at the point (x_{i+1}, y_j) may be written in terms its value, $u_{i,j}$ at the point (x_i, y_j) by means of a Taylor series expansion as follows:

$$u_{i+1,j} = u_{i,j} + \left.\frac{\partial u}{\partial x}\right|_{i,j} (\Delta x) + \left.\frac{\partial^2 u}{\partial x^2}\right|_{i,j} \frac{(\Delta x)^2}{2!} + \left.\frac{\partial^3 T}{\partial x^3}\right|_{i,j} \frac{(\Delta x)^3}{3!} + \ldots \tag{8.1}$$

This may be rearranged to make the first derivative of $u_{i,j}$ the subject as follows:

$$\begin{aligned} \left.\frac{\partial u}{\partial x}\right|_{i,j} &= \frac{u_{i+1,j} - u_{i,j}}{(\Delta x)} - \left.\frac{\partial^2 u}{\partial x^2}\right|_{i,j} \frac{(\Delta x)}{2!} - \left.\frac{\partial^3 u}{\partial x^3}\right|_{i,j} \frac{(\Delta x)^2}{3!} - \cdots \\ &= \frac{u_{i+1,j} - u_{i,j}}{(\Delta x)} + O(\Delta x) \end{aligned} \tag{8.2}$$

where $O(\Delta x)$ indicates the lowest order of the remaining terms. An estimate of the first derivative is obtained by truncating the series and ignoring terms of $O(\Delta x)$ and higher. That is,

$$\left.\frac{\partial u}{\partial x}\right|_{i,j} \approx \frac{u_{i+1,j} - u_{i,j}}{(\Delta x)} \tag{8.3}$$

Equation (8.3) is called a *first order, forward difference* approximation for the first derivative because it ignores terms of order one and greater and is written in terms of the value of u at a forward node, $u_{i+1,j}$.

If we choose $(\Delta x) < 0$, the value of $u_{i-1,j}$ may be obtained in terms of $u_{i,j}$ using a Taylor series as in Eq. (8.1):

$$u_{i-1,j} = u_{i,j} - \left.\frac{\partial u}{\partial x}\right|_{i,j}(\Delta x) + \left.\frac{\partial^2 u}{\partial x^2}\right|_{i,j}\frac{(\Delta x)^2}{2!} - \left.\frac{\partial^3 u}{\partial x^3}\right|_{i,j}\frac{(\Delta x)^3}{3!} + \cdots \tag{8.4}$$

Rearranging this expression, and ignoring terms of order one and higher, we obtain the *first order, backward difference* approximation to the first derivative:

$$\left.\frac{\partial u}{\partial x}\right|_{i,j} \approx \frac{u_{i,j} - u_{i-1,j}}{(\Delta x)} \tag{8.5}$$

In order to obtain more accurate approximations, subtract Eq. (8.4) from Eq. (8.1) to get

$$u_{i+1,j} - u_{i-1,j} = 2\left.\frac{\partial u}{\partial x}\right|_{i,j}(\Delta x) + 2\left.\frac{\partial^3 u}{\partial x^3}\right|_{i,j}\frac{(\Delta x)^3}{3!} + \cdots \tag{8.6}$$

From which we obtain,

$$\begin{aligned}\left.\frac{\partial u}{\partial x}\right|_{i,j} &= \frac{u_{i+1,j} - u_{i-1,j}}{2(\Delta x)} + \left.\frac{\partial^3 u}{\partial x^3}\right|_{i,j}\frac{(\Delta x)^2}{3!} + \cdots \\ &= \frac{u_{i+1,j} - u_{i-1,j}}{2(\Delta x)} + O(\Delta x^2)\end{aligned} \tag{8.7}$$

This leads to a *second order, central difference* approximation to the first derivative:

$$\left.\frac{\partial u}{\partial x}\right|_{i,j} \approx \frac{u_{i+1,j} - u_{i-1,j}}{2(\Delta x)} \tag{8.8}$$

To approximate second derivatives, add equations (8.1) and (8.4) to obtain

$$u_{i+1,j} + u_{i-1,j} = 2u_{i,j} + 2\frac{\partial^2 u}{\partial x^2}\bigg|_{i,j} \frac{(\Delta x)^2}{2!} + 2\frac{\partial^4 u}{\partial x^4}\bigg|_{i,j} \frac{(\Delta x)^4}{4!} + \cdots \qquad \textbf{(8.9)}$$

Rearranging we get,

$$\frac{\partial^2 u}{\partial x^2}\bigg|_{i,j} = \frac{u_{i+1,j} - 2u_{i,j} + u_{i-1,j}}{(\Delta x)^2} + O(\Delta x^2) \qquad \textbf{(8.10)}$$

From which a *second order, central difference* approximation is found for the second derivative:

$$\frac{\partial^2 u}{\partial x^2}\bigg|_{i,j} \approx \frac{u_{i+1,j} - 2u_{i,j} + u_{i-1,j}}{(\Delta x)^2} \qquad \textbf{(8.11)}$$

All the above steps can be repeated for the y-variable. The expressions corresponding to equations (8.3), (8.5), (8.8) and (8.11) are found to be:

- first order, forward difference approximation for the first derivative

$$\frac{\partial u}{\partial y}\bigg|_{i,j} \approx \frac{u_{i,j+1} - u_{i,j}}{(\Delta y)} \qquad \textbf{(8.12)}$$

- first order, backward difference approximation for the first derivative

$$\frac{\partial u}{\partial y}\bigg|_{i,j} \approx \frac{u_{i,j} - u_{i,j-1}}{(\Delta y)} \qquad \textbf{(8.13)}$$

- second order, central difference approximation for the first derivative

$$\frac{\partial u}{\partial y}\bigg|_{i,j} \approx \frac{u_{i,j+1} - u_{i,j-1}}{2(\Delta y)} \qquad \textbf{(8.14)}$$

- second order, central difference approximation for the second derivative

$$\frac{\partial^2 u}{\partial y^2}\bigg|_{i,j} \approx \frac{u_{i,j+1} - 2u_{i,j} + u_{i,j-1}}{(\Delta y)^2} \qquad \textbf{(8.15)}$$

8.2 A Simple Example

As an illustration of the application of a finite difference technique we solve the one dimensional problem used as an example in Chapter 7. That is, find the solution $u(x)$ such that

$$\frac{d^2u}{dx^2} - 2 = 0 \quad \text{for } 1 \leq x \leq 4 \tag{8.16}$$

subject to the boundary conditions:

$$\begin{aligned} \frac{du}{dx} &= 3\,, \quad \text{at } x = 1 \\ u(x) &= 20 \ \text{ at } x = 4 \end{aligned} \tag{8.17}$$

Recall from Section 7.6 that this equation may be solved exactly by integrating twice. The solution is

$$u(x) = x^2 + x \tag{8.18}$$

Figure 8.26 shows the mesh to be used. It consists of four nodes, equally spaced along the x-axis at positions $x = 1, 2, 3$ and 4.

Figure 8.26: Coarse Mesh for Finite Difference Solution

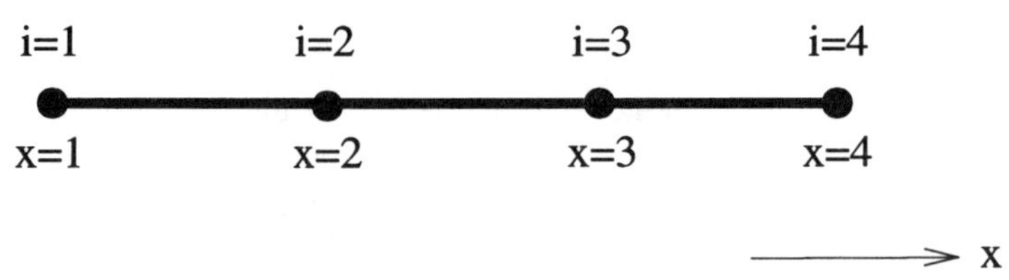

Using a second order, central difference for the second derivative (see Eq. (8.11)), Eq. (8.16) becomes,

$$\frac{u_{i+1} - 2u_i + u_{i-1}}{(\Delta x)^2} - 2 = 0 \tag{8.19}$$

or, after rearrangement,

$$u_{i+1} = 2(\Delta x)^2 + 2u_i - u_{i-1} \tag{8.20}$$

Since, for this example the internodal spacing is one, $(\Delta x) = 1$, and so this becomes:

$$u_{i+1} = 2 + 2u_i - u_{i-1} \tag{8.21}$$

Using a first order, forward difference approximation (see Eq. (8.3)) for du/dx, the boundary condition of Eq. (8.17) becomes:

$$\frac{u_2 - u_1}{(\Delta x)} = 3$$
$$\text{That is, } u_2 = 3 + u_1 \text{ as } \Delta x = 1 \tag{8.22}$$

Setting $i = 2$ in Eq. (8.21) we get:

$$u_3 = 2 + 2u_2 - u_1 \tag{8.23}$$

Substituting Eq. (8.22) we have

$$\begin{aligned} u_3 &= 2 + 2(3 + u_1) - u_1 \\ &= 8 + u_1 \end{aligned} \tag{8.24}$$

Setting $i = 3$ in Eq. (8.21) we get:

$$u_4 = 2 + 2u_3 - u_2 \tag{8.25}$$

Substituting equations (8.22) and (8.24) we have

$$\begin{aligned} u_4 &= 2 + 2(8 + u_1) - (3 + u_1) \\ &= 15 + u_1 \end{aligned} \tag{8.26}$$

But $u_4 = u(4) = 20$ by the boundary condition of Eq. (8.17). Therefore Eq. (8.26) becomes:

$$
\begin{aligned}
20 &= 15 + u_1 \\
u_1 &= 5
\end{aligned}
\tag{8.27}
$$

The values u_2 and u_3 can now be found by substituting $u_1 = 5$ in equations (8.22) and (8.24) respectively. We get $u_2 = 8$ and $u_3 = 13$. Our solution is therefore

$$
\begin{aligned}
u_1 &= 5 \qquad (3) \\
u_2 &= 8 \qquad (6) \\
u_3 &= 13 \quad (12) \\
u_4 &= 20 \quad (20)
\end{aligned}
\tag{8.28}
$$

where the exact solution is shown in parentheses.

A better solution is found by refining the mesh. Let's repeat the example with the mesh shown in Figure 8.27.

Figure 8.27: Fine Mesh for Finite Difference Solution

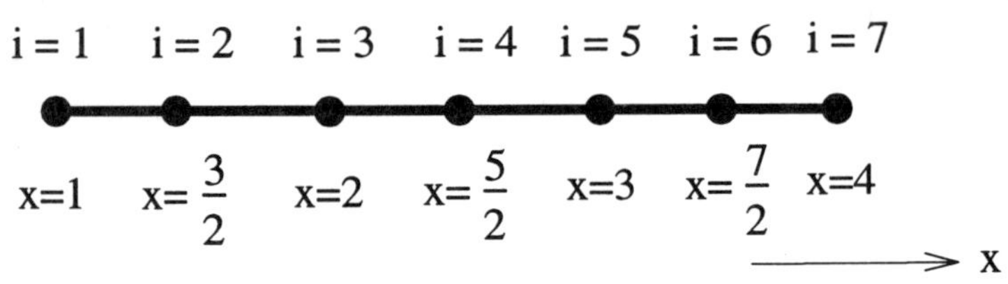

For this mesh, $(\Delta x) = 0.5$. Equation (8.20) therefore becomes:

$$
\begin{aligned}
u_{i+1} &= 2(0.5)^2 + 2u_i - u_{i-1} \\
&= 0.5 + 2u_i - u_{i-1}
\end{aligned}
\tag{8.29}
$$

The boundary condition of Eq. (8.22) becomes

$$
u_2 = 1.5 + u_1 \tag{8.30}
$$

Setting $i = 2$ in Eq. (8.29) we get

$$u_3 = 0.5 + 2u_2 - u_1 \tag{8.31}$$

Substituting Eq. (8.30) in the above we obtain

$$\begin{aligned} u_3 &= 0.5 + 2(1.5 + u_1) - u_1 \\ &= 3.5 + u_1 \end{aligned} \tag{8.32}$$

Continuing in this fashion we get the following equations:

$$\begin{aligned} u_4 &= 6 + u_1 \\ u_5 &= 9 + u_1 \\ u_6 &= 12.5 + u_1 \\ u_7 &= 16.5 + u_1 \end{aligned} \tag{8.33}$$

But $u_7 = u(4) = 20$ by the boundary condition of Eq. (8.17) and so from the last of Eq. (8.33) we have

$$\begin{aligned} 20 &= 16.5 + u_1 \\ u_1 &= 20 - 16.5 \\ &= 3.5 \end{aligned} \tag{8.34}$$

This may be substituted in (8.33) to find the following solution:

$$\begin{aligned} u_1 &= 3.5 && (3) \\ u_2 &= 5 && (3.75) \\ u_3 &= 7 && (6) \\ u_4 &= 9.5 && (8.75) \\ u_5 &= 12.5 && (12) \\ u_6 &= 16 && (15.75) \\ u_7 &= 20 && (20) \end{aligned} \tag{8.35}$$

where again the figures in parentheses indicate the exact solution. Comparing the above with the solution obtained earlier, it is clear that the refined mesh improves the solution accuracy.

8.3 The Diffusion Equation

Our use of finite difference methods is restricted to the one dimensional diffusion equation:

$$\frac{\partial T}{\partial t} = \alpha \frac{\partial^2 T}{\partial z^2} \tag{8.36}$$

where α is the diffusivity and T the temperature. In Chapter 9, the energy equation, which has a similar form to Eq. (8.36) is solved using finite difference techniques to give the temperature field through the cavity thickness.

Finite difference methods for diffusion equations fall into two broad categories: explicit or implicit methods. In the following sections these two classes of solution methods are discussed.

8.3.1 Explicit Methods

An explicit solution scheme arises when the spatial derivatives are evaluated at the old time step. Before considering this in more detail, we modify our notation slightly. Rather than denote the value of $T(z,t)$ at node (i,j) by $T_{i,j}$ as before, we use subscripts for the spatial node position and superscripts for the temporal nodes. Thus the value of $T(z,t)$ at node i in the z-direction and time step n will be denoted T_i^n.

To develop an explicit finite difference scheme for Eq. (8.36), the time derivative is replaced with a first order forward difference approximation. That is,

$$\frac{\partial T}{\partial t} \approx \frac{T_i^{n+1} - T_i^n}{\Delta t} \tag{8.37}$$

The spatial derivative is approximated with a second order central difference:

$$\frac{\partial^2 T}{\partial z^2} \approx \frac{T_{i+1}^n - 2T_i^n + T_{i-1}^n}{(\Delta z)^2} \tag{8.38}$$

Using these approximations, Eq. (8.36) may be written:

$$\frac{T_i^{n+1} - T_i^n}{\Delta t} = \alpha \left(\frac{T_{i+1}^n - 2T_i^n + T_{i-1}^n}{(\Delta z)^2} \right) \tag{8.39}$$

This may be rearranged to give the value of T at the new time $n+1$ as:

$$T_i^{n+1} = \frac{\alpha \Delta t}{(\Delta z)^2}\left(T_{i+1}^n - 2T_i^n + T_{i-1}^n\right) + T_i^n \\ = M\left(T_{i+1}^n - 2T_i^n + T_{i-1}^n\right) + T_i^n \tag{8.40}$$

where $M = \alpha\Delta t/(\Delta z)^2$. Using Eq. (8.40) the value at the new time is found explicitly from the values at the old time step.

Stability Criteria

Explicit schemes are quite fast but suffer from the fact that the time step Δt must be sufficiently small. For stable solution of (8.40), the following relation [AW] must be satisfied:

$$M = \frac{\alpha \Delta t}{(\Delta z)^2} \leq \frac{1}{2} \tag{8.41}$$

This condition means that time steps cannot be arbitrarily large.

8.3.2 Implicit Methods

In an implicit scheme, the time derivative is approximated in the same way as for the explicit method but the spatial derivative is approximated by a second order, central difference formula at the new time. Hence instead of Eq. (8.38) we use:

$$\frac{\partial^2 T}{\partial z^2} \approx \frac{T_{i+1}^{n+1} - 2T_i^{n+1} + T_{i-1}^{n+1}}{(\Delta z)^2} \tag{8.42}$$

Equation (8.36) therefore becomes

$$\frac{T_i^{n+1} - T_i^n}{\Delta t} = \alpha\left(\frac{T_{i+1}^{n+1} - 2T_i^{n+1} + T_{i-1}^{n+1}}{(\Delta z)^2}\right) \tag{8.43}$$

After rearrangement we obtain:

$$(1 + 2M)T_i^{n+1} - M\left(T_{i+1}^{n+1}\right) - M\left(T_{i-1}^{n+1}\right) = T_i^n \tag{8.44}$$

When written for each node, Eq. (8.44) yields a set of linear equations which may be expressed in matrix form as follows:

$$[A]\{T\}^{n+1} = \{T\}^n \tag{8.45}$$

Here, the matrix $[A]$ is a square matrix of known terms involving M and constants, $\{T\}^{n+1}$ is a matrix of temperatures at the new time and $\{T\}^n$ is a matrix of known temperatures at the old time. Equation (8.45) may then be solved for the temperatures at the new time.

Implicit methods do not suffer from the stability requirement of Eq. (8.41). They are stable for any size time step. However, if the time step is too large they can converge to the wrong solution albeit in a stable manner. This advantage must be balanced against the need to solve the matrix Eq. (8.45) at each time step. Typically when there are many nodes, a numerical scheme such as Gauss-Seidel iteration is required.

In Moldflow software, both implicit and explicit methods are used. The choice being dependent on the time step involved. Details may be found in Section 9.4.

Chapter 9
Numerical Solution for the Filling Phase

In this Chapter numerical solution of the simplified equations governing the filling phase is outlined. Solution of the equations is done in three main stages: calculation of the pressure field, calculation of the temperature field and finally, advancement of the flow front. Finite element methods are used for the pressure calculation while temperature calculations use finite difference methods. Advancement of the flow front uses a control-volume approach.

9.1 Overview of Solution Process

The governing equations for the filling phase were derived in Chapter 5. They are:

$$\frac{\partial}{\partial x}\left(S_2\frac{\partial p}{\partial x}\right)+\frac{\partial}{\partial y}\left(S_2\frac{\partial p}{\partial y}\right)=0$$
$$\text{where } S_2=\int_0^h \frac{z^2}{\eta}dz \tag{9.1}$$

where,

$$S = \int_0^h \frac{z^2}{\eta} dz \tag{9.2}$$

and

$$\rho c_p \left(\frac{\partial T}{\partial t} + v_x \frac{\partial T}{\partial x} + v_y \frac{\partial T}{\partial y} \right) = \eta \dot{\gamma}^2 + k \frac{\partial^2 T}{\partial z^2}$$
$$\text{where } \dot{\gamma}^2 = \left(\frac{\partial v_x}{\partial z} \right)^2 + \left(\frac{\partial v_y}{\partial z} \right)^2 \tag{9.3}$$

Equations (9.1) and (9.3) are to be solved subject to the following boundary conditions:

i. The pressure is zero at the flow front, that is,

$$p = 0 \text{ on flow front} \tag{9.4}$$

ii. The pressure or flow rate is defined at the point(s) where the melt is injected. That is,

$$p = p_{in} \text{ or } q = q_{in} \text{ at the injection point(s).} \tag{9.5}$$

iii. The normal pressure gradient is zero at any impermeable boundary. That is at a boundary we have

$$\frac{\partial p}{\partial n} = 0 \tag{9.6}$$

iv. The temperature at the cavity wall or at some point interior to the mold wall is defined. That is,

$$T = T_w \text{ at } z = h \text{ or } z = h + \delta \tag{9.7}$$

v. The temperature gradient in the z-direction is zero at the cavity centre line. That is,

$$\frac{\partial T}{\partial z} = 0 \text{ at } z = 0 \tag{9.8}$$

vi. The melt temperature at the point(s) of injection is defined. That is,

$$T = T_{in} \text{ at the injection point(s)} \tag{9.9}$$

Due to the dependence of the fluidity S_2 on viscosity, which in turn depends on both temperature and shear rate, equations (9.1) and (9.3) must be solved simultaneously. However, for numerical solution, the equations are decoupled by using small time increments. At a particular time, the temperature is assumed constant and the pressure field is calculated by assuming a value of viscosity at that temperature. Provided that the time increments are sufficiently small this decoupling provides satisfactory results.

The sequence of the solution process is as follows:

1. Calculate S_2. If this is the start of an analysis, a nominal value of viscosity at the melt temperature is used. If it is not the first step of an analysis, shear rate and temperature data from a previous step may be used to calculate S_2. With the fluidity known, we solve Eq. (9.1) for pressure subject to the boundary conditions (9.4), (9.5) and (9.6). After calculating the pressure field, it is possible to determine the velocity using equations (5.44) and (5.45). A new value of shear rate may then be calculated using the expression in Eq. (9.3). Assuming constant temperature, the viscosity is updated using this value of shear rate. This viscosity value is then used to calculate the fluidity S_2. Equation (9.1) is now solved again and the entire process repeated. We continue to iterate in this way until, on two consecutive iterations, the change in pressure is less than some defined tolerance.
2. After the pressure calculation has converged, the current value of velocities v_x and v_y, the shear rate $\dot{\gamma}$ and the viscosity η are used in Eq. (9.3) to calculate the convective and viscous heating terms. Solution of (9.3) is therefore reduced to a conduction problem with convection and viscous heating treated as source terms. The conduction calculations are performed with finite difference methods and give the temperature field. With temperature now known, an updated viscosity value is calculated. This is then used to calculate the flow into each control volume on the flow front.
3. Knowing the flow rate into each control volume it is possible to predict which of these will fill in the next time increment. The flow front is then advanced accordingly.

Steps 1, 2 and 3 above are repeated until the mold is filled.

It should be noted that compressibility effects are also accounted for in the pressure field calculation described in item 1 above. This will be described in Section 9.6.

In the remainder of this Chapter, we look at these steps in more detail.

9.2 Finite Element Solution for the Pressure Field

The cavity geometry is discretized into triangular elements. Although it is possible to use other shapes such as quadrilateral elements, triangular elements can "fill" the complex geometries encountered in injection molding and allows automatic meshing. Automatic mesh generation dramatically reduces modeling time.

Runners are modeled with elements of circular cross-section and will be discussed later.

9.2.1 Finite Element Formulation for Triangular Elements

The cavity model is meshed with triangular elements. Linear interpolation functions are used for the pressure within an element. Hence the pressure at any point (x, y) within an element is approximately given by:

$$\tilde{p}^{(e)}(x, y) = \sum_{i=1}^{3} N_i p_i \tag{9.10}$$

Where N_i are the linear shape functions and p_i are the values of pressure at node i.

Note that:

$$\begin{aligned} N_i &= 1 \text{ at node } i \\ N_i &= 0 \text{ at the other nodes.} \end{aligned}$$

The shape functions have a particularly simple form if we work in area coordinates [CR]. Consider the triangular element shown in Figure (9.28).

Figure 9.28: Area Coordinates for Triangular Elements

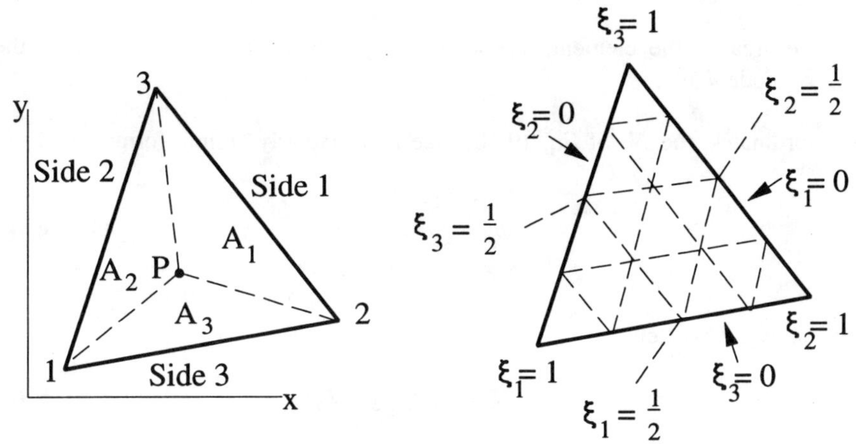

Area coordinates are based on the fact that a point $P(x, y)$ in a triangle divides the region into three areas defined by straight lines from the vertices to the point P. Area coordinates ξ_1, ξ_2 and ξ_3 are defined as follows:

$$\xi_1 = \frac{A_1}{A} \qquad \xi_2 = \frac{A_2}{A} \qquad \xi_3 = \frac{A_3}{A} \tag{9.11}$$

where A is the area of the triangle. Obviously:

$$\xi_1 + \xi_2 + \xi_3 = 1 \tag{9.12}$$

The relationship between area coordinates and the global x-y system are as follows:

$$\begin{Bmatrix} 1 \\ x \\ y \end{Bmatrix} = \begin{bmatrix} 1 & 1 & 1 \\ x_1 & x_2 & x_3 \\ y_1 & y_2 & y_3 \end{bmatrix} \begin{Bmatrix} \xi_1 \\ \xi_2 \\ \xi_3 \end{Bmatrix} \tag{9.13}$$

$$\begin{Bmatrix} \xi_1 \\ \xi_2 \\ \xi_3 \end{Bmatrix} = \frac{1}{2A} \begin{bmatrix} x_2y_3 - x_3y_2 & y_{23} & x_{32} \\ x_3y_1 - x_1y_3 & y_{31} & x_{13} \\ x_1y_2 - x_2y_1 & y_{12} & x_{21} \end{bmatrix} \begin{Bmatrix} 1 \\ x \\ y \end{Bmatrix} \tag{9.14}$$

where A is the area of the element, $x_{ij} = x_i - x_j$ and $y_{ij} = y_i - y_j$ with x_i the x-coordinate of node i etc.

Using area coordinates, the N_i of Eq. (9.10) take a particularly simple form:

$$\begin{aligned} N_1 &= \xi_1 \\ N_2 &= \xi_2 \\ N_3 &= \xi_3 \end{aligned} \tag{9.15}$$

Hence (9.10) may be written

$$\tilde{p}(\xi_1, \xi_2, \xi_3) = \xi_1 p_1 + \xi_2 p_2 + \xi_3 p_3 \tag{9.16}$$

Before continuing we introduce a more concise notation for the pressure equation. We write

$$\frac{\partial}{\partial x}\left(S_2 \frac{\partial p}{\partial x}\right) + \frac{\partial}{\partial y}\left(S_2 \frac{\partial p}{\partial y}\right) = (\nabla \cdot S_2 \nabla p) \tag{9.17}$$

where ∇ is taken to be the two dimensional divergence and gradient operator. Hence the pressure equation (9.1) may be written

$$(\nabla \cdot S_2 \nabla p) = 0 \tag{9.18}$$

To derive the element equations we use Galerkin's method (see Section 7.4.2). The residual $R(\tilde{p})$ is given by

$$R(\tilde{p}) = (\nabla \cdot \nabla S_2 \tilde{p}) \tag{9.19}$$

The Galerkin method requires that

$$\begin{aligned} 0 &= \int_A N_i R(\tilde{p}) dA \\ &= \int_A N_i (\nabla \cdot S_2 \nabla \tilde{p}) dA \end{aligned} \tag{9.20}$$

The divergence of the product of a scalar, a and vector $\vec{x}$ can be expressed as

$$\nabla \cdot (a\vec{x}) = a\nabla \cdot \vec{x} + \vec{x} \cdot \nabla a \tag{9.21}$$

Applying this identity to (9.20) gives

$$\begin{aligned} 0 &= \int_A N_i \nabla \cdot (S_2 \nabla \tilde{p}) \, dA \\ &= \int_A \nabla \cdot (N_i S_2 \nabla \tilde{p}) \, dA - \int_A S_2 \nabla \tilde{p} \cdot \nabla N_i \, dA \end{aligned} \tag{9.22}$$

Applying the divergence theorem in the plane (see Section 4.1.2) to the first term on the right hand side of Eq. (9.22) gives

$$0 = \int_\Sigma N_i S_2 \nabla \tilde{p} \cdot \vec{n} d\Sigma - \int_A S_2 \nabla \tilde{p} \cdot \nabla N_i dA \quad i = 1, 2, 3 \tag{9.23}$$

where Σ is the boundary of the element.

That is

$$\int_A S_2 \nabla \tilde{p} \cdot \nabla N_i dA = \int_\Sigma N_i S_2 \nabla \tilde{p} \cdot \vec{n} d\Sigma \quad i = 1, 2, 3 \tag{9.24}$$

In order to use the above we need expressions for $\nabla \tilde{p}$ and ∇N_i. First note that by Eq. (9.16) we can write

$$\begin{aligned} \frac{\partial \tilde{p}}{\partial x} &= \frac{\partial}{\partial x}(\xi_1 p_1 + \xi_2 p_2 + \xi_3 p_3) \\ \frac{\partial \tilde{p}}{\partial y} &= \frac{\partial}{\partial y}(\xi_1 p_1 + \xi_2 p_2 + \xi_3 p_3) \end{aligned} \tag{9.25}$$

Differentiating (9.14) we obtain,

$$\begin{aligned}\frac{\partial \xi_1}{\partial x} &= \frac{y_{23}}{2A} & \frac{\partial \xi_1}{\partial y} &= \frac{x_{32}}{2A}\\ \frac{\partial \xi_2}{\partial x} &= \frac{y_{31}}{2A} & \frac{\partial \xi_2}{\partial y} &= \frac{x_{13}}{2A}\\ \frac{\partial \xi_3}{\partial x} &= \frac{y_{12}}{2A} & \frac{\partial \xi_3}{\partial y} &= \frac{x_{21}}{2A}\end{aligned} \tag{9.26}$$

Using these relationships we have,

$$\begin{aligned}\frac{\partial \tilde{p}}{\partial x} &= \frac{\partial}{\partial x}(\xi_1 p_1 + \xi_2 p_2 + \xi_3 p_3)\\ &= \frac{1}{2A}(y_{23}p_1 + y_{31}p_2 + y_{12}p_3)\end{aligned} \tag{9.27}$$

and

$$\begin{aligned}\frac{\partial \tilde{p}}{\partial y} &= \frac{\partial}{\partial y}(\xi_1 p_1 + \xi_2 p_2 + \xi_3 p_3)\\ &= \frac{1}{2A}(x_{32}p_1 + x_{13}p_2 + x_{21}p_3)\end{aligned} \tag{9.28}$$

Hence

$$\nabla \tilde{p} = \frac{1}{2A}(y_{23}p_1 + y_{31}p_2 + y_{12}p_3)\hat{e}_1 + \frac{1}{2A}(x_{32}p_1 + x_{13}p_2 + x_{21}p_3)\hat{e}_2 \tag{9.29}$$

Also, using equations (9.15) and (9.14) we obtain expressions for ∇N_i:

$$\begin{aligned}\nabla N_i &= \nabla \xi_i\\ &= \frac{\partial \xi_i}{\partial x}\hat{e}_1 + \frac{\partial \xi_i}{\partial y}\hat{e}_2 \quad i = 1, 2, 3\end{aligned} \tag{9.30}$$

Equations (9.29) and (9.30) can now be used in Eq. (9.24) to give three algebraic equations. For example, for $i = 1$, (9.24) becomes

$$\int_A S_2 \nabla \tilde{p} \cdot \nabla N_1 dA = \int_\Sigma N_1 S_2 \nabla \tilde{p} \cdot \vec{n} d\Sigma \tag{9.31}$$

We now evaluate the LHS and RHS of Eq. (9.31). To do this we assume that S_2 is constant over the element at a given time and temperature. Substituting equations (9.29) and (9.30) into the LHS of (9.31) gives:

$$\begin{aligned}\int_A S_2 \nabla\tilde{p}.\nabla N_1 dA &= S_2 \int_A \nabla\tilde{p}\cdot\nabla N_1 dA \\ &= \frac{S_2}{2A}\int_A \Big\{(y_{23}p_1 + y_{31}p_2 + y_{12}p_3)\frac{\partial\xi_1}{\partial x} \\ &\quad + (x_{32}p_1 + x_{13}p_2 + x_{21}p_3)\frac{\partial\xi_1}{\partial y}\Big\}\, dA \\ &= \frac{S_2}{4A^2}\int_A \Big\{(y_{23}^2 + x_{32}^2)p_1 + (y_{31}y_{23} + x_{13}x_{32})p_2 \\ &\quad + (y_{12}y_{23} + x_{21}x_{32})p_3\Big\}\, dA \\ &= \frac{S_2}{4A}\Big\{(y_{23}^2 + x_{32}^2)p_1 + (y_{31}y_{23} + x_{13}x_{32})p_2 \\ &\quad + (y_{12}y_{23} + x_{21}x_{32})p_3\Big\}\end{aligned} \tag{9.32}$$

We now turn to the problem of evaluating the RHS of Eq. (9.31).

Figure 9.29: Geometry of Triangular Element

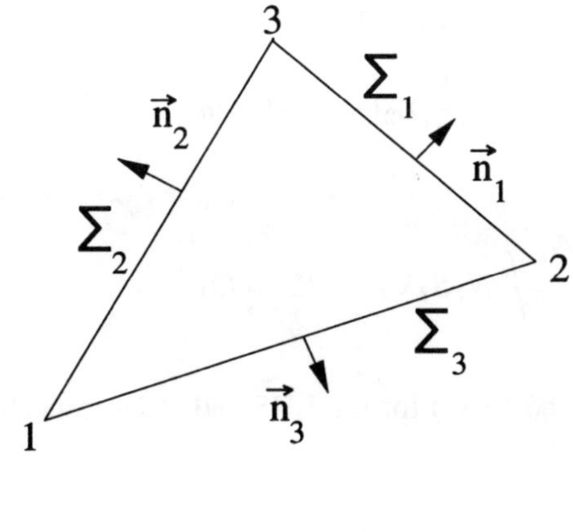

In Figure (9.29), the $\vec{n}_i$ are unit outward normals to the element sides and the element boundary Σ is composed of the three lines Σ_1, Σ_2 and Σ_3. Because $N_1 = \xi_1$ is identically zero on Σ_1, the only contribution to the RHS of Eq. (9.31) is along sides Σ_2 and Σ_3. Substituting equations (9.29) and (9.30) in the RHS of (9.31) we get

$$\begin{aligned}\int_{\Sigma} N_1 S_2 \nabla\tilde{p}\cdot\vec{n}\,d\Sigma &= \int_{\Sigma_2\cup\Sigma_3} \xi_1 S_2 \frac{\partial P}{\partial n}\,d\Sigma \\ &= S_2\frac{\partial P}{\partial n_2}\int_{\Sigma_2}\xi_1 d\Sigma + S_2\frac{\partial P}{\partial n_3}\int_{\Sigma_3}\xi_1 d\Sigma \\ &= S_2\frac{\partial P}{\partial n_2}|\Sigma_2|\int_0^1 x\,dx + S_2\frac{\partial P}{\partial n_3}|\Sigma_3|\int_0^1 (1-x)\,dx \\ &= \frac{1}{2}S_2\frac{\partial P}{\partial n_2}|\Sigma_2| + \frac{1}{2}S_2\frac{\partial P}{\partial n_3}|\Sigma_3| \\ &= \frac{1}{2}q_2|\Sigma_2| + \frac{1}{2}q_3|\Sigma_3|\end{aligned} \tag{9.33}$$

where $q_2 = S_2\partial P/\partial n_2$, $q_3 = S_2\partial P/\partial n_3$, $|\Sigma_2|$ is the length of side 2 and $|\Sigma_3|$ is the length of side 3.

The quantities q_2 and q_3 may be identified as flow rates per unit length along sides 2 and 3 respectively. Hence the sum of $1/2|\Sigma_2|q_2$ and $1/2|\Sigma_3|q_3$ may be defined as the total flow rate at node 1, Q_1. We write

$$Q_1 = \frac{1}{2}|\Sigma_2|q_2 + \frac{1}{2}|\Sigma_3|q_3 \tag{9.34}$$

Hence (9.33) becomes:

$$\int_{\Sigma} N_1 S_2 \nabla\tilde{p}\cdot\vec{n}\,d\Sigma = Q_1 \tag{9.35}$$

Substituting equations (9.32) and (9.35) for the LHS and RHS respectively of Eq. (9.24) we obtain, for the case $i = 1$:

$$\frac{S_2}{4A}\left\{(y_{23}^2 + x_{32}^2)p_1 + (y_{31}y_{32} + x_{13}x_{32})p_2 + (y_{12}y_{32} + x_{21}x_{32})p_3\right\} = Q_1 \tag{9.36}$$

Setting $i = 2$ and $i = 3$ in Eq. (9.24) and carrying out similar calculations to the above we obtain the following two equations:

$$\frac{S_2}{4A}\{(y_{23}y_{31} + x_{32}x_{13})p_1 + (y_{31}^2 + x_{13}^2)p_2 + (y_{12}y_{31} + x_{21}x_{13})p_3\} = Q_2 \quad \textbf{(9.37)}$$

$$\frac{S_2}{4A}\{(y_{23}y_{12} + x_{32}x_{21})p_1 + (y_{31}y_{12} + x_{13}x_{21})p_2 + (y_{12}^2 + x_{21}^2)p_3\} = Q_3 \quad \textbf{(9.38)}$$

where

$$\begin{aligned} Q_2 &= \frac{1}{2}|\Sigma_3|q_3 + \frac{1}{2}|\Sigma_1|q_1 \\ Q_3 &= \frac{1}{2}|\Sigma_2|q_2 + \frac{1}{2}|\Sigma_1|q_1 \end{aligned} \quad \textbf{(9.39)}$$

and

$$q_1 = S_2\frac{\partial P}{\partial n_1} \quad \textbf{(9.40)}$$

Equations (9.36), (9.37) and (9.38) are called the *element equations*. They give the relationship between pressure and flow rate for any triangular element. For brevity and subsequent analysis, the element equations are usually written in matrix notation as follows:

$$\frac{S}{4A}\begin{bmatrix} (y_{23}^2 + x_{32}^2) & (y_{31}y_{23} + x_{13}x_{32}) & (y_{12}y_{23} + x_{21}x_{32}) \\ (y_{23}y_{31} + x_{32}x_{13}) & (y_{31}^2 + x_{13}^2) & (y_{12}y_{31} + x_{21}x_{13}) \\ (y_{23}y_{12} + x_{32}x_{21}) & (y_{31}y_{12} + x_{13}x_{21}) & (y_{12}^2 + x_{21}^2) \end{bmatrix}\begin{Bmatrix} p_1 \\ p_2 \\ p_3 \end{Bmatrix} = \begin{Bmatrix} Q_1 \\ Q_2 \\ Q_3 \end{Bmatrix} \quad \textbf{(9.41)}$$

The three by three matrix on the LHS is called the *element stiffness matrix*. Note that the element stiffness matrix is *symmetric*. Therefore to describe the element stiffness equations we really only require six terms.

Equation (9.41) may be further abbreviated to:

$$\begin{bmatrix} k_{11} & k_{12} & k_{13} \\ k_{21} & k_{22} & k_{23} \\ k_{31} & k_{32} & k_{33} \end{bmatrix} \begin{Bmatrix} p_1 \\ p_2 \\ p_3 \end{Bmatrix} = \begin{Bmatrix} Q_1 \\ Q_2 \\ Q_3 \end{Bmatrix} \quad \textbf{(9.42)}$$

where, $k_{11} = \frac{S}{4A}(y_{23}^2 + x_{32}^2)$, $k_{12} = \frac{S}{4A}(y_{31}y_{23} + x_{13}x_{23})$ etc. and, due to symmetry, $k_{ij} = k_{ji}$.

The element equations can be abbreviated further to:

$$[K]^e\{p\} = \{Q\} \quad \textbf{(9.43)}$$

where the e superscript indicates that these equations are for an element and not the total system. Later we will see that after assembly, the resulting system equations have the same structure as the element equations.

Equation (9.42) gives the relationship between pressure and flow rate for any triangular element. To model the entire cavity, the element equations are assembled to extend the relationship to the entire solution domain. We discuss this in a later section.

9.2.2 Finite Element Formulation for Runner Elements

In this section the element equations for runner elements are derived. These can be assembled with the triangular elements to form the system equations for the entire cavity and feed system.

The governing equation for pressure in runners of circular cross-section was derived in Chapter 5. It is

$$\frac{\partial}{\partial x}\left(r^+ S_1 \frac{\partial p}{\partial x}\right) = 0 \quad \textbf{(9.44)}$$

where r^+ defines the radial coordinate of the interface between the frozen layer and the melt, and

$$S_1 = \frac{1}{2r^+}\int_0^{r^+} \frac{r^3}{\eta}dr \quad \textbf{(9.45)}$$

At this stage we allow r^+ and S_1 to vary with x.

Figure (9.30) illustrates a runner element of circular cross-section. Each runner element is defined by two nodes located at x_1 and x_2. The pressure at nodes 1 and 2 is denoted by p_1 and p_2 respectively. Flow rate into the element at node 1 is denoted by q_1 and flow rate out at node 2 is q_2.

Figure 9.30: Runner Element of Circular Cross-Section

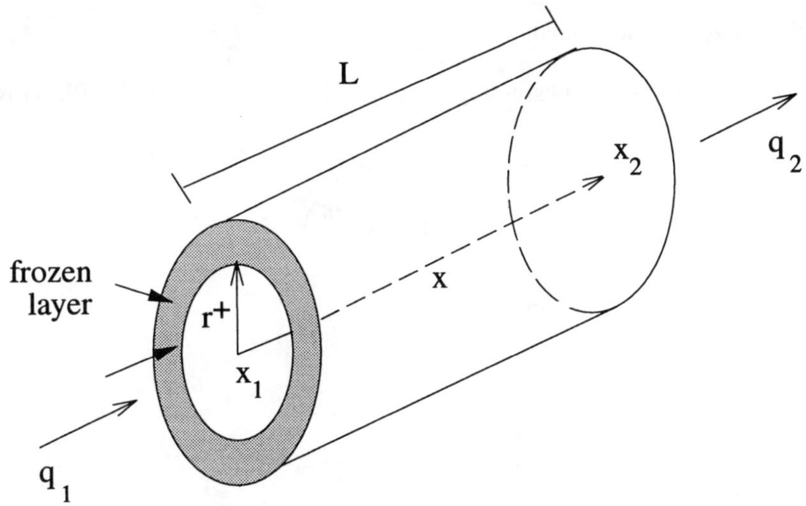

We assume that the pressure in the element may be represented as a linear function of the nodal pressures. Hence we use linear shape functions as discussed in Section 7.3. The approximate pressure solution is therefore written as:

$$\tilde{p} = \left(\frac{x_2 - x}{x_2 - x_1}\right)p_1 + \left(\frac{x - x_1}{x_2 - x_1}\right)p_2 \qquad \textbf{(9.46)}$$

That is,

$$\tilde{p} = N_1 p_1 + N_2 p_2 \qquad \textbf{(9.47)}$$

where

$$
\begin{aligned}
N_1 &= \frac{x_2 - x}{x_2 - x_1} = \frac{x_2 - x}{L} \\
N_2 &= \frac{x - x_1}{x_2 - x_1} = \frac{x - x_1}{L}
\end{aligned}
\tag{9.48}
$$

and L is the length of the element.

The residual $R(x)$ created by substituting the approximate solution in Eq. (9.44) is

$$
R(x) = \frac{\partial}{\partial x}\left(r^+ S_1 \frac{\partial \tilde{p}}{\partial x}\right) \tag{9.49}
$$

Using the Galerkin method we set

$$
\int_{x_1}^{x_2} R(x) N_i dx = 0 \quad i = 1, 2 \tag{9.50}
$$

Substituting Eq. (9.49) in (9.50) we obtain

$$
\begin{aligned}
0 &= \int_{x_1}^{x_2} R(x) N_i dx \\
&= \int_{x_1}^{x_2} \frac{\partial}{\partial x}\left(r^+ S_1 \frac{\partial \tilde{p}}{\partial x}\right) N_i dx \qquad i = 1, 2 \\
&= \left(r^+ S_1 \frac{\partial \tilde{p}}{\partial x} N_i\right)\Bigg|_{x_1}^{x_2} - \int_{x_1}^{x_2} r^+ S_1 \frac{\partial \tilde{p}}{\partial x}\frac{\partial N_i}{\partial x} dx \qquad i = 1, 2
\end{aligned}
\tag{9.51}
$$

This yields two equations, corresponding to $i = 1$ and $i = 2$:

$$
\begin{aligned}
0 &= \int_{x_1}^{x_2} R(x) N_1 dx \\
&= \left[r^+ S_1 \frac{\partial \tilde{p}}{\partial x} N_1 \right]_{x=x_1}^{x=x_2} - \int_{x_1}^{x_2} r^+ S_1 \frac{\partial \tilde{p}}{\partial x} \frac{\partial N_1}{\partial x} dx \\
&= -\left(r^+ S_1 \frac{\partial \tilde{p}}{\partial x} \right) \bigg|_{x=x_1} - \int_{x_1}^{x_2} r^+ S_1 \frac{\partial \tilde{p}}{\partial x} \frac{\partial N_1}{\partial x} dx \\
\text{and } 0 &= \int_{x_1}^{x_2} R(x) N_2 dx \\
&= \left[r^+ S_1 \frac{\partial \tilde{p}}{\partial x} N_2 \right]_{x=x_1}^{x=x_2} - \int_{x_1}^{x_2} r^+ S_1 \frac{\partial \tilde{p}}{\partial x} \frac{\partial N_2}{\partial x} dx \\
&= \left(r^+ S_1 \frac{\partial \tilde{p}}{\partial x} \right) \bigg|_{x=x_2} - \int_{x_1}^{x_2} r^+ S_1 \frac{\partial \tilde{p}}{\partial x} \frac{\partial N_2}{\partial x} dx
\end{aligned}
\qquad \textbf{(9.52)}
$$

Differentiating Eq. (9.48) we find

$$
\frac{\partial N_1}{\partial x} = -\frac{1}{L} \qquad \frac{\partial N_2}{\partial x} = \frac{1}{L} \qquad \textbf{(9.53)}
$$

and so from Eq. (9.47),

$$
\begin{aligned}
\frac{\partial \tilde{p}}{\partial x} &= \frac{\partial}{\partial x} \{ N_1 p_1 + N_2 p_2 \} \\
&= \frac{\partial N_1}{\partial x} p_1 + \frac{\partial N_2}{\partial x} p_2 \\
&= -\frac{1}{L}(p_1 - p_2)
\end{aligned}
\qquad \textbf{(9.54)}
$$

The first terms on the RHS of equations (9.52) may be identified as volumetric flow rates into and out of nodes 1 and 2 respectively. Using Eq. (9.54) we define flow rates, q_1 and q_2 as follows:

$$
\begin{aligned}
q_1 &= \left[r^+ S_1 \frac{\partial \tilde{p}}{\partial x} N_1 \right]_{x=x_1}^{x=x_2} \\
&= -\left[\frac{r^+ S_1}{L} (p_1 - p_2) \right]_{x=x_1} \\
q_2 &= \left[r^+ S_1 \frac{\partial \tilde{p}}{\partial x} N_2 \right]_{x=x_1}^{x=x_2} \\
&= \left[\frac{r^+ S_1}{L} (p_1 - p_2) \right]_{x=x_2}
\end{aligned}
\tag{9.55}
$$

Hence (9.52) becomes

$$
\begin{aligned}
0 &= q_1 - \int_{x_1}^{x_2} r^+ S_1 \frac{\partial \tilde{p}}{\partial x} \frac{\partial N_1}{\partial x} dx \\
0 &= q_2 - \int_{x_1}^{x_2} r^+ S_1 \frac{\partial \tilde{p}}{\partial x} \frac{\partial N_2}{\partial x} dx
\end{aligned}
\tag{9.56}
$$

Consider now the second term on the RHS of Eq. (9.56). Using (9.53) and (9.54) we get

$$
\begin{aligned}
\int_{x_1}^{x_2} r^+ S_1 \frac{\partial \tilde{p}}{\partial x} \frac{\partial N_1}{\partial x} dx &= \int_{x_1}^{x_2} \frac{r^+ S_1}{L^2} (p_1 - p_2) dx \\
&= \frac{r^+ S_1}{L} (p_1 - p_2) \\
\int_{x_1}^{x_2} r^+ S_1 \frac{\partial \tilde{p}}{\partial x} \frac{\partial N_2}{\partial x} dx &= -\int_{x_1}^{x_2} \frac{r^+ S_1}{L^2} (p_1 - p_2) dx \\
&= -\frac{r^+ S_1}{L} (p_1 - p_2)
\end{aligned}
\tag{9.57}
$$

where, in the last step, we have assumed that r^+ and S_1 are constant over the element.

Using equations (9.55) and (9.57), we can write Eq. (9.52) in the form:

$$
\begin{aligned}
0 &= \int_{x_1}^{x_2} R(x) N_1 dx \\
&= q_1 - \frac{r^+ S_1}{L}(p_1 - p_2) \\
0 &= \int_{x_1}^{x_2} R(x) N_2 dx \\
&= q_2 + \frac{r^+ S_1}{L}(p_1 - p_2)
\end{aligned}
\tag{9.58}
$$

This may be written in matrix notation as:

$$
\frac{r^+ S_1}{L}\begin{bmatrix} 1 & -1 \\ -1 & 1 \end{bmatrix}\begin{Bmatrix} p_1 \\ p_2 \end{Bmatrix} = \begin{Bmatrix} q_1 \\ q_2 \end{Bmatrix} \tag{9.59}
$$

9.2.3 Assembly of Element Equations

In the last section the element equations (9.42), were derived for a triangular element. These equations give the relationship between nodal pressures and flow rates for a single unattached element. While these equations have identical form for all elements, the numerical values will vary from element to element. The finite element model for the cavity comprises many elements. To model the entire cavity we therefore have to *assemble* the individual element equations.

For simplicity we will consider the assembly of a simple cavity that consists of just two triangular elements. A superscript denotes the element to which the relevant terms apply.

Figure (9.31) shows the two elements to be assembled. Element 1 is defined by nodes 1, 2 and 4. Element 2 is defined by nodes 2, 3 and 4.

Figure 9.31: Assembly of Two Element Model

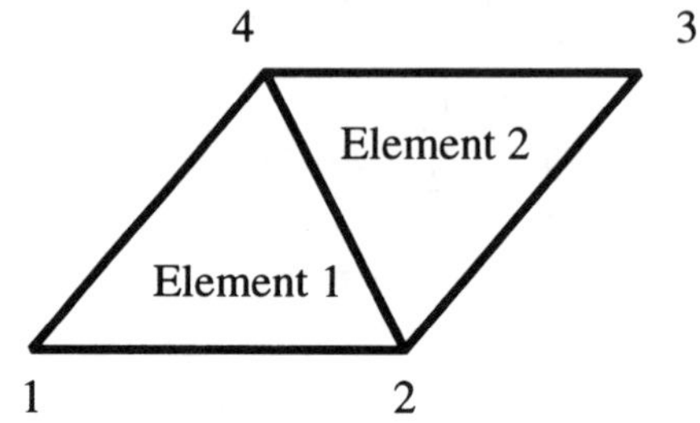

The element equations for elements one and two are, respectively,

$$\begin{bmatrix} k_{11}^{(1)} & k_{12}^{(1)} & k_{13}^{(1)} \\ k_{21}^{(1)} & k_{22}^{(1)} & k_{23}^{(1)} \\ k_{31}^{(1)} & k_{32}^{(1)} & k_{33}^{(1)} \end{bmatrix} \begin{Bmatrix} p_1^{(1)} \\ p_2^{(1)} \\ p_4^{(1)} \end{Bmatrix} = \begin{Bmatrix} Q_1^{(1)} \\ Q_2^{(1)} \\ Q_4^{(1)} \end{Bmatrix} \qquad \textbf{(9.60)}$$

$$\begin{bmatrix} k_{11}^{(2)} & k_{12}^{(2)} & k_{13}^{(2)} \\ k_{21}^{(2)} & k_{22}^{(2)} & k_{23}^{(2)} \\ k_{31}^{(2)} & k_{32}^{(2)} & k_{33}^{(2)} \end{bmatrix} \begin{Bmatrix} p_2^{(2)} \\ p_3^{(2)} \\ p_4^{(2)} \end{Bmatrix} = \begin{Bmatrix} Q_2^{(2)} \\ Q_3^{(2)} \\ Q_4^{(2)} \end{Bmatrix} \qquad \textbf{(9.61)}$$

Note that each element equation involves a 3 by 3 matrix. The first step in the assembly procedure is to expand the element equations so that 4 by 4 matrices can be used. The reason for this will become clear later. The element equations for element 1 can be written using 4 by 4 matrices as follows:

$$\begin{bmatrix} k_{11}^{(1)} & k_{12}^{(1)} & 0 & k_{13}^{(1)} \\ k_{21}^{(1)} & k_{22}^{(1)} & 0 & k_{23}^{(1)} \\ 0 & 0 & 0 & 0 \\ k_{31}^{(1)} & k_{32}^{(1)} & 0 & k_{33}^{(1)} \end{bmatrix} \begin{Bmatrix} p_1 \\ p_2 \\ p_3 \\ p_4 \end{Bmatrix} = \begin{Bmatrix} Q_1^{(1)} \\ Q_2^{(1)} \\ 0 \\ Q_4^{(1)} \end{Bmatrix} \qquad \textbf{(9.62)}$$

The equations for element 2 can be written:

$$\begin{bmatrix} 0 & 0 & 0 & 0 \\ 0 & k_{11}^{(2)} & k_{12}^{(2)} & k_{13}^{(2)} \\ 0 & k_{21}^{(2)} & k_{22}^{(2)} & k_{23}^{(2)} \\ 0 & k_{31}^{(2)} & k_{32}^{(2)} & k_{33}^{(2)} \end{bmatrix} \begin{Bmatrix} p_1 \\ p_2 \\ p_3 \\ p_4 \end{Bmatrix} = \begin{Bmatrix} 0 \\ Q_2^{(2)} \\ Q_3^{(2)} \\ Q_4^{(2)} \end{Bmatrix} \qquad \textbf{(9.63)}$$

The assembled equations are found by simply adding together equations (9.62) and (9.63) to get:

$$\begin{bmatrix} k_{11}^{(1)} & k_{12}^{(1)} & 0 & k_{13}^{(1)} \\ k_{21}^{(1)} & k_{22}^{(1)} + k_{11}^{(2)} & k_{12}^{(2)} & k_{23}^{(1)} + k_{13}^{(2)} \\ 0 & k_{21}^{(2)} & k_{22}^{(2)} & k_{23}^{(2)} \\ k_{13}^{(1)} & k_{32}^{(1)} + k_{31}^{(2)} & k_{32}^{(2)} & k_{33}^{(2)} \end{bmatrix} \begin{Bmatrix} p_1 \\ p_2 \\ p_3 \\ p_4 \end{Bmatrix} = \begin{Bmatrix} Q_1^{(1)} \\ Q_2^{(1)} + Q_2^{(2)} \\ Q_3^{(2)} \\ Q_4^{(1)} + Q_4^{(2)} \end{Bmatrix} \tag{9.64}$$

Equation (9.64) is the matrix form of the *system equations* for the simple cavity modeled with two elements. The four by four matrix on the LHS is called the *global stiffness matrix*.

Of course injection molded parts are not so simple. Finite element models for flow analysis frequently contain thousands of nodes. Consequently, the global stiffness matrix can be very large.

It should be noted that the stiffness matrix gets larger as the flow front progresses. Initially the stiffness matrix may only involve a few elements. As the melt fills the cavity, the filled and partially filled elements are added to the stiffness matrix.

Before discussing application of boundary conditions we remark that for filled nodes common to two or more elements, the total nodal flow rate is zero. This must be so to conserve mass. For example in Eq. (9.64), if node 2 is filled, we would have $Q_2^{(1)} = -Q_2^{(2)}$ and so the total flow rate at node 2, $Q_2^{(1)} + Q_2^{(2)} = 0$. In summary, the flow into a filled node must be equal to the flow out. This means that as the flow front progresses, the majority of terms on the RHS of the system equations are zero. Total flow rates are non-zero at the injection nodes and for nodes that are on the flow front but not yet filled.

9.2.4 Application of Boundary Conditions

Having assembled the element equations, the next step is to enforce the boundary conditions. These were given in Section 9.1.

The relevant boundary conditions are:

- nodes on the flow front have assigned pressure zero and
- the pressure or flow rate is known at the injection node(s).

For nodes on the flow front, the boundary condition that pressure is zero is enforced by deleting the rows and columns of the stiffness matrix that correspond to the node number(s). For example, if nodes 3, 4 and 7 are on the flow front we would delete the rows and columns numbered 3, 4 and 7 from the stiffness matrix.

For the injection nodes the known pressures or flow rates are substituted in the system equations.

With the boundary conditions set, the system equations may be solved to determine the nodal pressures. These are then substituted back in the system equations to get the corresponding nodal flow rates.

As an example, in Figure (9.32), we indicate the injection node for our simple two element cavity. The flow rate into the injection node (node 1) is given and is $\bar{Q}_1^{(1)}$, say. Let's suppose that nodes 2 and 4 have filled and so the flow front position is in element two. Therefore, $p_3 = 0$, since node 3 is on the flow front. This means that row 3 and column 3 are deleted from the equations.

Figure 9.32: Boundary Conditions for Simple Cavity

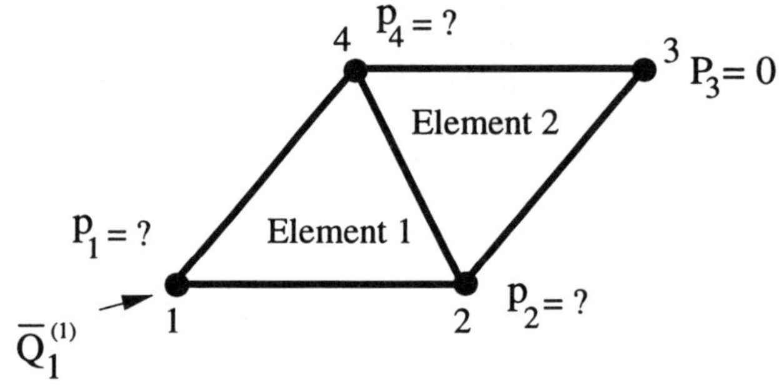

The unknowns are p_1, p_2, p_4 and Q_3. Note that because nodes 2 and 4 are filled, $Q_2^{(1)} + Q_2^{(2)} = 0$ and $Q_4^{(1)} + Q_4^{(2)} = 0$.

Hence equations (9.64) become:

$$\begin{bmatrix} k_{11}^{(1)} & k_{12}^{(1)} & k_{13}^{(1)} \\ k_{21}^{(1)} & k_{22}^{(1)} + k_{11}^{(2)} & k_{23}^{(1)} + k_{13}^{(2)} \\ k_{13}^{(1)} & k_{32}^{(1)} + k_{31}^{(2)} & k_{33}^{(2)} \end{bmatrix} \begin{Bmatrix} p_1 \\ p_2 \\ p_4 \end{Bmatrix} = \begin{Bmatrix} \overline{Q}_1^{(1)} \\ 0 \\ 0 \end{Bmatrix} \tag{9.65}$$

The boundary conditions have reduced the number of equations to be solved

9.2.5 Solution of the System Equations

The set of equations (9.65) can be solved to give p_1, p_2 and p_4. With these pressures known, the flow rate into node 3, $Q_3^{(2)}$ can be found by back substitution in Eq. (9.64).

As mentioned earlier, the global stiffness matrix is typically quite large and solution of the system equations must be done numerically. In Moldflow software, we use Gauss Seidel iteration with successive under-relaxation to solve for nodal pressures.

Gauss Seidel iteration may be explained as follows. For a simple system with a stiffness matrix that is 3x3:

$$\begin{bmatrix} k_{11} & k_{12} & k_{13} \\ k_{21} & k_{22} & k_{23} \\ k_{31} & k_{32} & k_{33} \end{bmatrix} \begin{Bmatrix} p_1 \\ p_2 \\ p_3 \end{Bmatrix} = \begin{Bmatrix} Q_1 \\ Q_2 \\ Q_3 \end{Bmatrix} \tag{9.66}$$

we start with an estimate of the pressures. If we denote the new value of pressure obtained after one iteration by p', we get:

$$\begin{aligned} p_1' &= (Q_1 - k_{12}p_2 - k_{13}p_3)/k_{11} \\ p_2' &= (Q_2 - k_{21}p_1' - k_{23}p_3)/k_{22} \\ p_3' &= (Q_3 - k_{31}p_1' - k_{32}p_2')/k_{33} \end{aligned} \tag{9.67}$$

For the next iteration, p' is replaced by p'', and so on. Iterations stop when the difference between pressures calculated on two consecutive iterations is less than a specified tolerance.

Convergence can be sped up by using an over-relaxation factor, β. In this case, rather than use the latest pressure p' we use $p_i + \beta(p_i' - p_i)$. The value of β is chosen between 1 and 2.

For a system consisting of N nodes, the method of successive over-relaxation takes the form:

$$p_i' = p_i + \frac{\beta}{k_{ii}} \left(Q_i - \sum_{j=1}^{i-1} k_{ij} p_j' - \sum_{j=1}^{N} k_{ij} p_j \right) \tag{9.68}$$

The quantity in brackets in the above equation may be viewed as an imbalance in flow rate. The solution technique tries to minimize this imbalance.

Solution of large sets of equations is a topic in itself. The interested reader should consult some of the many books on this subject [HY], [YG].

9.3 Finite Difference Solution of the Energy Equation for the Cavity

The energy equation, after simplification, is

$$\rho c_p \left(\underbrace{\frac{\partial T}{\partial t}}_{\textit{time derivative}} + \underbrace{v_x \frac{\partial T}{\partial x} + v_y \frac{\partial T}{\partial y}}_{\textit{convection}} \right) = \underbrace{\eta \dot{\gamma}^2}_{\textit{dissipation}} + \underbrace{k \frac{\partial^2 T}{\partial z^2}}_{\textit{conduction}} \qquad \textbf{(9.69)}$$

and this is to be solved subject to the boundary conditions:

- The temperature at the cavity wall or at some point interior to the mold wall is defined. That is,

$$T = T_w \text{ at } z = h \text{ or } z = h + \delta \qquad \textbf{(9.70)}$$

- The temperature gradient in the z-direction is zero at the cavity centre line. That is,

$$\frac{\partial T}{\partial z} = 0 \text{ at } z = 0 \qquad \textbf{(9.71)}$$

- The melt temperature at the point(s) of injection is defined. That is,

$$T = T_{in} \text{ at the injection point(s)} \qquad \textbf{(9.72)}$$

9.3.1 Finite Difference Grid Definition

For numerical solution, it is necessary to discretize Eq. (9.69) in both the time t, and spatial dimension z.

Time discretization is closely linked with the growing scheme used for advancement of the mesh (see later in Section 9.5).

The spatial discretization is as shown in Figure (9.33).

Figure 9.33: Discretization of the Cavity for Thermal Calculations

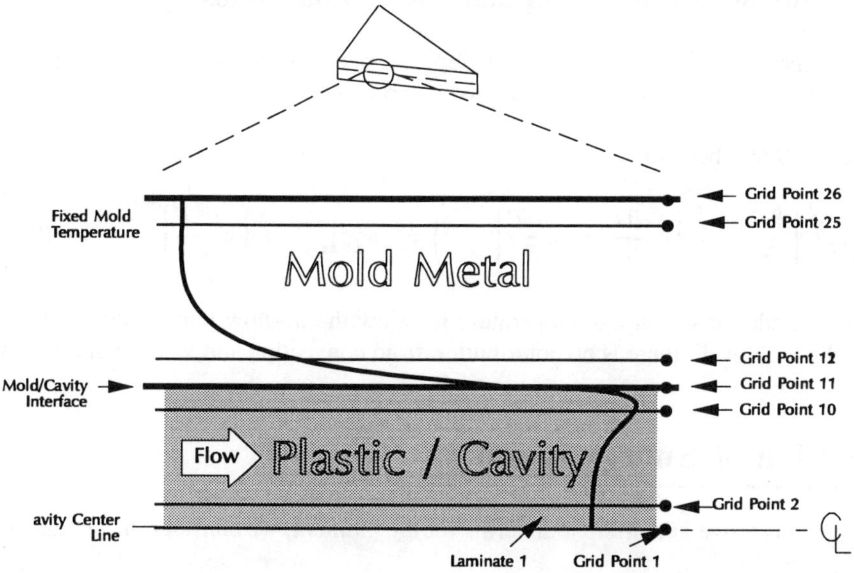

Grid Points and Laminates across Thickness of an Element

In the figure, each element has 26 points defined for thermal calculations. Of these, 11 are located in the cavity, with grid point 1 at the cavity centre line and grid point 11 at the cavity/mold wall interface. Grid points 12 to 25 are located in the mold wall. This allows the cavity/mold interface temperature to vary.

Grid points in the cavity are equally spaced. Grid points in the mold wall are also uniformly spaced but the spacing is determined by the grid point ratio, R, which is defined by:

$$R = \frac{\textit{spacing between consecutive grid points in the metal}}{\text{spacing between consecutive grid points in the cavity}} \tag{9.73}$$

This means that the distance between consecutive grid points in the metal is R times that of the distance between consecutive points in the cavity.

9.3.2 Treatment of Convection and Dissipation Terms

For finite difference solution, in any time step, the convection and viscous dissipation terms from the previous time step are treated as source terms.

That is, Eq. (9.69) becomes:

$$\rho c_p \left(\left[\frac{\partial T}{\partial t} \right]_i^n + \left[v_x \frac{\partial T}{\partial x} + v_y \frac{\partial T}{\partial y} \right]_i^{n-1} \right) = \left[\eta \dot{\gamma}^2 \right]_i^{n-1} + \left[k \frac{\partial^2 T}{\partial z^2} \right]_i^n \tag{9.74}$$

Of course for nodes at which the temperature is below the no-flow temperature or nodes located in the mold wall, there is no contribution from convection and viscous dissipation.

9.4 Solution of Energy Equation

Ignoring the convective and dissipation terms for the moment, we consider the conduction part of Eq. (9.69):

$$\rho c_p \left(\frac{\partial T}{\partial t} \right) = k \frac{\partial^2 T}{\partial z^2} \tag{9.75}$$

This can be rearranged to give:

$$\frac{\partial T}{\partial t} = \alpha \frac{\partial^2 T}{\partial z^2} \tag{9.76}$$

where $\alpha = k/(\rho c_p)$ is the thermal diffusivity.

Recall from Section 1 that for stable solution of Eq. (9.76) using an explicit finite difference scheme, we require

$$\begin{aligned} M &= \frac{\alpha \Delta t}{(\Delta z)^2} \\ &\leq \frac{1}{2} \end{aligned} \tag{9.77}$$

To minimize computation time, it is possible to use a mix of explicit and implicit solution schemes. For example, in a particular time step Δt^n, the stability criterion is checked. There are several possibilities:

1. The stability requirement is met, that is, $M < 1/2$. In this case an *explicit* finite difference method is used for solution.
2. The stability requirement is not met, with $1/2 \leq M < 2$. In this case we perform $M/0.5$ explicit time steps.
3. The stability requirement is not met, with $2 \leq M < 10$. In this case we perform one step using an implicit finite difference scheme.
4. The stability requirement is not met, with $M \geq 10$. In this case we perform $M/10$ steps using an implicit finite difference scheme for each.

9.4.1 Explicit Finite Difference Method

For this case we use a first order forward difference approximation for the time derivative and a second order, central difference for the spatial term (see Section 8.3.1). That is,

$$\begin{aligned} \frac{\partial T}{\partial t} &\approx \frac{T_i^{n+1} - T_i^n}{t_i^{n+1} - t^n} \\ &= \frac{T_i^{n+1} - T_i^n}{\Delta t} \\ \frac{\partial^2 T}{\partial z^2} &\approx \frac{T_{i+1}^n - 2T_i^n + T_{i-1}^n}{(\Delta z)^2} \end{aligned} \tag{9.78}$$

Using these relationships, Eq. (9.76) can be written:

$$\frac{T_i^{n+1} - T_i^n}{\Delta t} = \frac{\alpha\left(T_{i+1}^n - 2T_i^n + T_{i-1}^n\right)}{(\Delta z)^2} \tag{9.79}$$

Rearranging we obtain,

$$T_i^{n+1} = s\left(T_{i+1}^n - 2T_i^n + T_{i-1}^n\right) + T_i^n \tag{9.80}$$

If we now add in the convection and dissipation terms, we get:

$$T_i^{n+1} = M\left(T_{i+1}^n - 2T_i^n + T_{i-1}^n\right) + T_i^n - (\Delta t)\left[v_x \frac{\partial T}{\partial x} + v_y \frac{\partial T}{\partial y}\right]_i^n + \frac{\Delta t}{\rho c_p}\left[\eta\dot{\gamma}^2\right]_i^n \tag{9.81}$$

This equation can then be solved for T_i^{n+1}.

9.4.2 Implicit Finite Difference Method

An implicit solution technique has the advantage of being unconditionally stable for any time step size (see Section 8.3.2). However, it is more compute intensive than the explicit scheme used above.

As in the explicit case, if the equation is solved at time step t^{n+1} then the convective and viscous dissipation terms from the previous time step may be added as source terms to the conduction equation.

Using a two point forward difference expression for the time derivative and a three point central difference expression for the spatial derivative, we have;

$$\begin{aligned} \frac{\partial T}{\partial t} &\approx \frac{T_i^{n+1} - T_i^n}{\Delta t} \\ \frac{\partial^2 T}{\partial z^2} &\approx \frac{T_{i+1}^{n+1} - 2T_i^{n+1} + T_{i-1}^{n+1}}{(\Delta z)^2} \end{aligned} \tag{9.82}$$

Substituting these expressions in Eq. (9.74) and rearranging gives:

$$-MT_{i+1}^{n+1} + (1+2M)T_i^{n+1} - MT_{i-1}^{n+1} = T_i^n - \Delta t\left[v_x \frac{\partial T}{\partial x} + v_y \frac{\partial T}{\partial y}\right]_i^n + \frac{\Delta t}{\rho c_p}\left[\eta\dot{\gamma}^2\right]_i^n \tag{9.83}$$

where

$$M = \frac{k\Delta t}{\rho c_p (\Delta z)^2} \tag{9.84}$$

Equation (9.83) holds for grid point i at any node. The unknown values of T at the new time, t^{n+1} are found by writing Eq. (9.83) for every filled node in the model. These equations can then be arranged in matrix form as follows:

$$[M]\{T^{n+1}\} = \{b\} \tag{9.85}$$

where $[M]$ is a square matrix, $\{T^{n+1}\}$ is a column matrix of unknown values of T at the new time and $\{b\}$ is a column matrix of the old values of T and source terms associated with the convective and viscous heating terms. The matrix equation is readily solved for $\{T^{n+1}\}$ using Gauss-Seidel iteration.

9.4.3 Numerical Solution of the Energy Equation for Runners

The energy equation for runners was established in Section 5.5.3 and is,

$$\rho c_p \left(\frac{\partial T}{\partial t} + v_x \frac{\partial T}{\partial x} \right) = \eta \dot{\gamma}^2 + k \frac{1}{r} \frac{\partial}{\partial r} \left(r \frac{\partial T}{\partial r} \right) \tag{9.86}$$

Convection and viscous dissipation are evaluated at a previous time step and, for the current time step, are treated as source terms. This is identical to the procedure described in Section 9.3 and will not be discussed further.

The variation of temperature with time due to conduction rate is given by

$$\frac{\partial T}{\partial t} = \frac{k}{\rho c_p} \frac{1}{r} \frac{\partial}{\partial r} \left(r \frac{\partial T}{\partial r} \right) \tag{9.87}$$

Differentiating the RHS gives

$$\frac{\partial T}{\partial t} = \frac{k}{\rho c_p} \left(\frac{1}{r} \left(\frac{\partial T}{\partial r} \right) + \frac{\partial^2 T}{\partial r^2} \right) \tag{9.88}$$

A solution for T may be found using either implicit or explicit methods.

9.5 Advancement of the Flow Front

The flow front is advanced using a control volume method. In this scheme each node is assigned a volume. These control volumes are defined by the polygonal region formed by linking the mid-sides of a triangular element to its centroid by a straight line segment. Figure (9.34) illustrates a section of the mesh and the associated control volumes.

Figure 9.34: Control Volumes and Associated Nodes

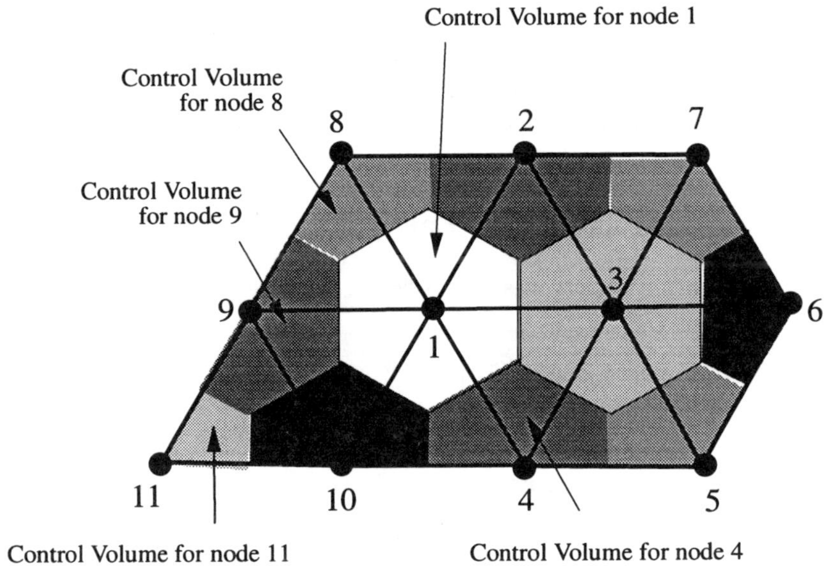

After solving for pressure, the flow rates into each node on the flow front can be calculated. Since the time step is known, the node can be tested to see if it is full. Once the node is filled, the flow front is advanced by incorporating all nodes connected to the last node to fill into the flow front. We illustrate this below.

9.5.1 Nodal Growth of Flow Front

An illustrated example of the stages of the nodal growth mechanism follows. For simplicity, resistance to flow is assumed the same in all directions.

1. Node 1 - the gate node - has filled. The first estimate of the flow front is the line joining all nodes that are attached to node 1 (nodes 2, 7, 5, 3, 6 and 4). This is the initial flow front. Note that - being equidistant from node 1 - these nodes are all equally likely to fill at the same time.

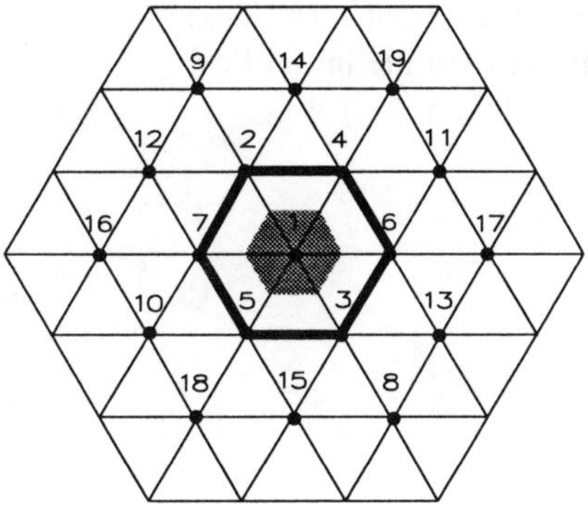

2. Assume that node 2, on the initial flow front, is the first to fill. The flow front is now expanded by all the nodes connected to node number 2. However - as flow always takes the path of least resistance, the shortest distance in this instance - the nodes connected to node 1 fill before those connected to node 2. In any event, the nodes connected to node 2 receive very little flow at this stage. With nodes 7, 5, 3, 6 and 4 still filling, one of these is the next to fill

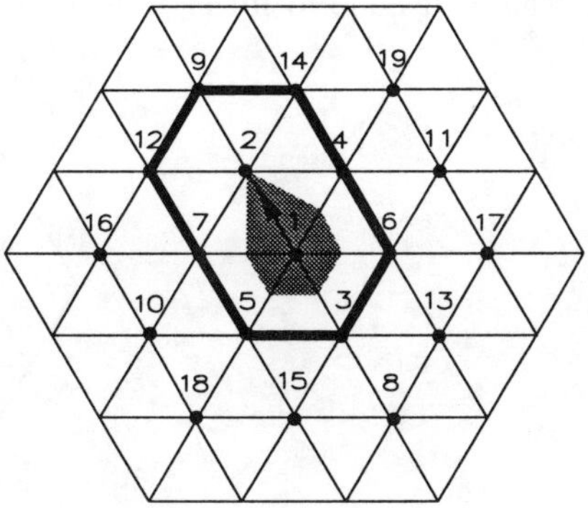

3. Assume that node 5 is the next to fill. Its adjoining nodes are now added to the flow front. The next node to fill - still on the initial flow front - is either 7, 3, 6 or 4.

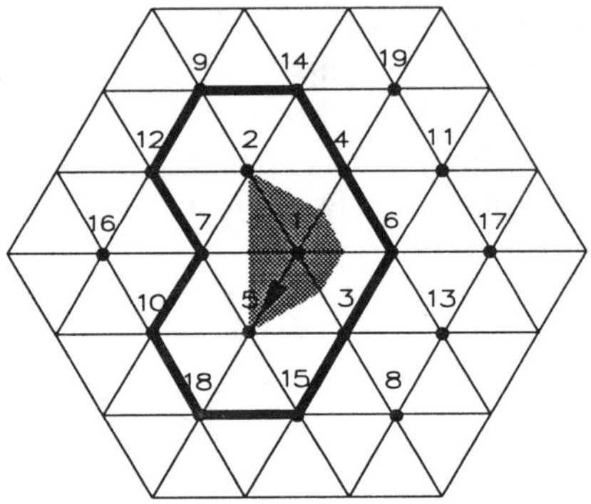

4. Assuming that node 6 is the next to fill, the flow front is expanded as shown. The next node to fill is either 7, 3 or 4

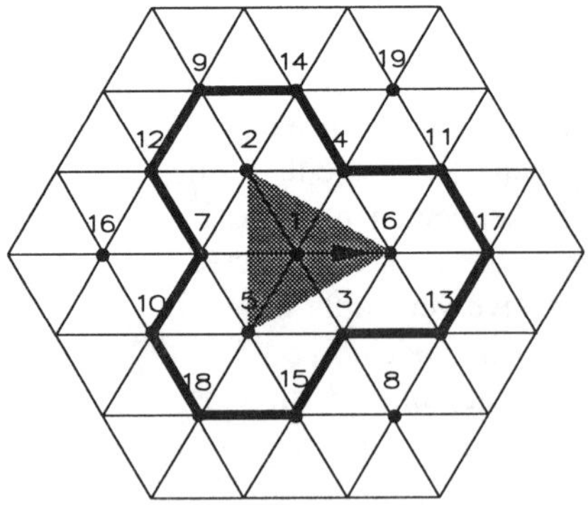

5. Assuming that node 7 has filled, the next node to fill is either 3 or 4

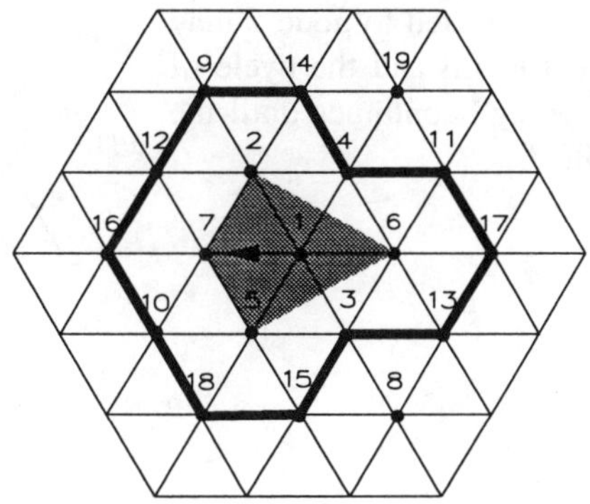

6. Assume that node 3 is the next to fill.

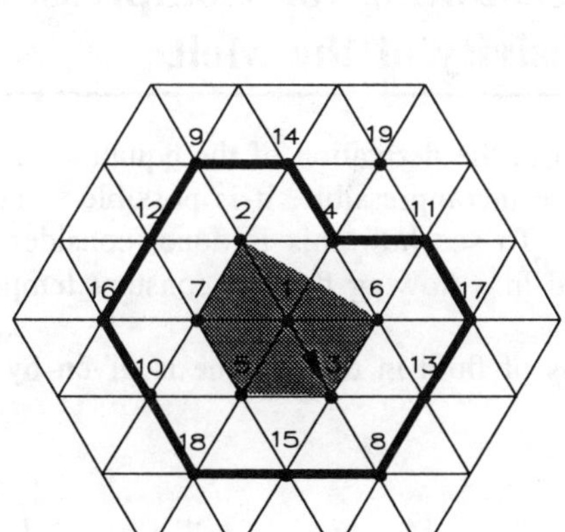

7. After the last node (node 4) on the initial flow front is filled, the flow front is as shown. One of the nodes connected to node 2 may be the next to fill and the cycle of "node hopping" continues until the part is filled.

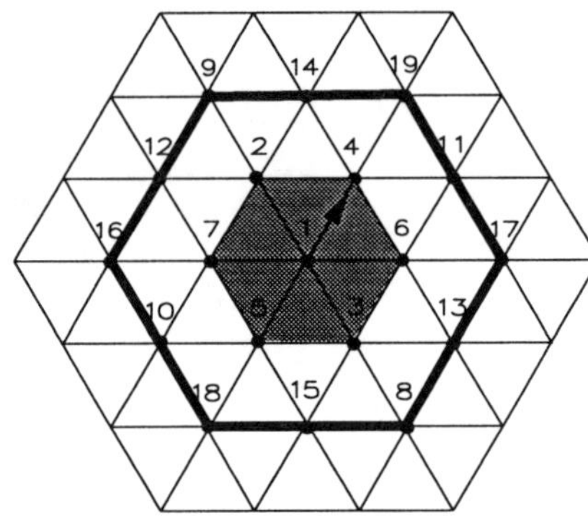

9.6 Accounting for Compressibility and Expansivity of the Melt

Although in the derivation of the equations governing the filling phase we assumed the melt to be incompressible, it is possible to account for compressibility in the pressure solution. To see how this is done, consider a material volume V, with boundary S, immersed in a flowing fluid at constant temperature.

The mass of fluid in the volume is given by

$$m = \int_V \rho dV \tag{9.89}$$

Using conservation of mass, the rate of change of mass in the volume must be zero. That is:

$$\begin{aligned} 0 &= \frac{\partial}{\partial t} \int_V \rho dV \\ &= \int_V \left(\frac{\partial \rho}{\partial t} + \nabla \cdot \rho \vec{v} \right) dV \\ &\text{(using the transport theorem)} \\ &= \int_V \frac{\partial \rho}{\partial t} dV + \int_S \rho \vec{v} \cdot \vec{n} dS \\ &\text{(using the divergence theorem)} \end{aligned} \tag{9.90}$$

where $\vec{n}$ is a unit outward vector on S.

Noting that $\partial \rho / \partial t = (\partial \rho / \partial p)(\partial p / \partial t)$, we have:

$$\begin{aligned} \int_V \frac{\partial \rho}{\partial t} dV &= \int_V \frac{\partial \rho}{\partial p} \frac{\partial p}{\partial t} dV \\ &= - \int_V \rho \kappa \frac{\partial p}{\partial t} dV \end{aligned} \tag{9.91}$$

where κ, is the isothermal compressibility coefficient (see Section 6.2).

We now associate a *control volume* with each node. These control volumes are used for compressibility calculations and advancing the flow front (see Section 9.5).

Figure (9.35) shows a section of the finite element mesh used to describe the mold cavity. The triangular elements are shown in solid black lines. Control volumes are shown in dotted lines. Control volume boundaries for a particular node are defined by lines that go through the mid side and centroids of connected elements.

Figure 9.35: Definition of Control Volumes

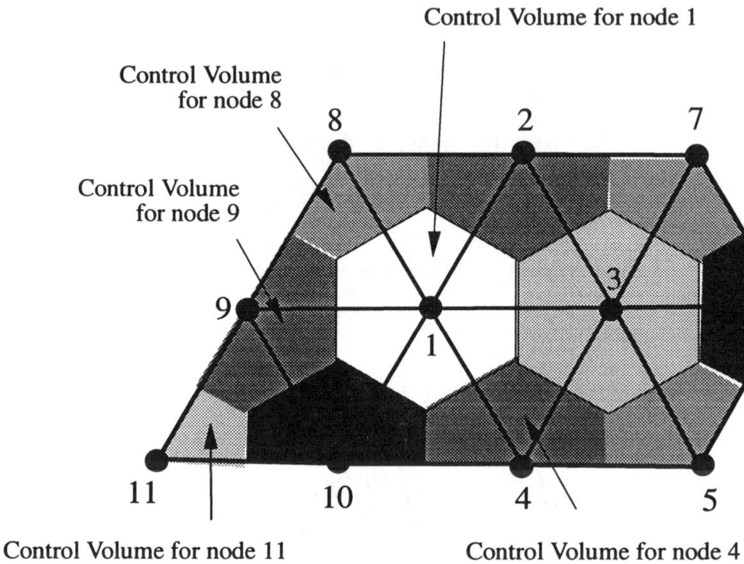

If we now assume that the pressure in each control volume is constant, Eq. (9.91) becomes

$$\int_V \frac{\partial \rho}{\partial t} dV = -\int_V \rho\kappa \frac{\partial p}{\partial t} dV = -\rho\kappa V \frac{\partial p}{\partial t} \qquad \textbf{(9.92)}$$

Hence Eq. (9.90) may be written:

$$\begin{aligned}0 &= \frac{d}{dt}\int_V dV \\ &= \int_V \frac{\partial \rho}{\partial t} dV + \int_S \rho \vec{v}\cdot\vec{n} dS \\ &\text{(using the divergence theorem)} \\ &= -\rho\kappa\frac{\partial p}{\partial t}V + \rho\int_S \vec{v}\cdot\vec{n} dS\end{aligned} \tag{9.93}$$

The last term in the above expression may be identified as the difference between the rate of flow out of V and the rate of flow in. For an incompressible fluid this last term is zero. Hence the rate of flow q_c, due to compressibility, is given by:

$$q_c = -\rho\kappa V\frac{\partial p}{\partial t} \tag{9.94}$$

In the simulation software, the control volume is in fact divided into a number of laminates by the finite difference grid used for thermal calculations. Hence temperature can vary across the control volume thickness. This allows density to be determined at each laminate. A flow rate due to compressibility can be calculated for each laminate using (9.94) above. The total flow rate due to compressibility at node number i, $q_(c_i)$ is then defined as the sum of the laminate flow rates.

Compressibility can dramatically affect the delivery of melt to the cavity. At the start of fill, the relatively large volume of fluid ahead of the ram is not highly compressed. As the ram starts to move forward this material is compressed but mass flow rate into the cavity can be far less than expected.

The compressibility model described above is valid under isothermal conditions. Temperature variation can be taken into account with the expansivity of the material. The melt expansivity, β is defined as

$$\beta = \frac{1}{\widehat{V}}\left(\frac{\partial \widehat{V}}{\partial T}\right)_p \tag{9.95}$$

Using the chain rule this may be written:

$$\left(\frac{\partial V}{\partial t}\right)_p = \beta V \left(\frac{\partial T}{\partial t}\right)_p \tag{9.96}$$

The LHS may be considered as a flow rate , q_e due to temperature change in a control volume. That is,

$$q_e = \beta V \left(\frac{\partial T}{\partial t}\right)_p \tag{9.97}$$

Just as for compressibility, these flow rates can be calculated for each laminate in the control volume. The total flow rate due to expansivity, q_{e_i} at the control volume associated with node i is then obtained by summing the flow rate for each laminate.

In practice, q_{c_i} and q_{c_i}) must be calculated by approximating the derivatives. Values of ρ, β and κ are obtained from the PVT data for the material. The derivatives are approximated using values on two consecutive time steps. That is,

$$\begin{aligned} q_c &= -\rho\kappa V \frac{\partial p}{\partial t} \\ &\approx -\rho\kappa \frac{p_{1+1} - p_i}{\Delta t} \\ q_e &= \rho\beta V \frac{\partial T}{\partial t} \\ &\approx \rho\beta V \frac{T_{1+1} - T_i}{\Delta t} \end{aligned} \tag{9.98}$$

The total flow rate for a node is incorporated in the pressure solution as follows. Whereas for the incompressible case we solved the system equations with all nodal flows zero (except the injection node(s) where the flow rate is known), we now solve with the values of $q_{c_i} + q_{e_i}$ on the RHS for all filled nodes. For the simple example in Section 9.2.4, the system equations become:

$$\begin{bmatrix} k_{11}^{(1)} & k_{12}^{(1)} & 0 & k_{13}^{(1)} \\ k_{21}^{(1)} & k_{22}^{(1)} + k_{11}^{(2)} & k_{12}^{(2)} & k_{23}^{(1)} + k_{13}^{(2)} \\ 0 & k_{21}^{(2)} & k_{22}^{(2)} & k_{23}^{(2)} \\ k_{13}^{(1)} & k_{32}^{(1)} + k_{31}^{(2)} & k_{32}^{(2)} & k_{33}^{(2)} \end{bmatrix} \begin{Bmatrix} p_1 \\ p_2 \\ p_3 \\ p_4 \end{Bmatrix} = \begin{Bmatrix} Q_1^{(1)} \\ q_{c_2} + q_{e_2} \\ Q_3^{(2)} \\ q_{c_4} + q_{e_4} \end{Bmatrix} \tag{9.99}$$

Chapter 10
Numerical Solution of the Packing Phase

This Chapter describes the numerical solution of the packing phase. Unlike the filling phase, the effects of compressibility and expansivity are included at all nodes in the cavity and feed system during packing.

Because the cavity is volumetrically filled prior to packing, there is no need for flow front advancement. As the nodal growth scheme used in the filling phase determines time steps, it is necessary to prescribe time steps in the packing phase.

Boundary conditions for packing are different to those used in filling. In particular the boundary condition that pressure is zero on the flow front is no longer applicable.

10.1 Overview of the Solution Process

The governing equations for the packing phase were derived in Chapter 6. For the cavity, they are:

$$\frac{\partial}{\partial x}\left(S_2\frac{\partial p}{\partial x}\right)+\frac{\partial}{\partial y}\left(S_2\frac{\partial p}{\partial y}\right)= \overline{\kappa}H\frac{\partial p}{\partial t}-\int_0^H\frac{\beta}{\rho c_p}\left(\eta\dot{\gamma}^2+k\frac{\partial^2 T}{\partial z^2}\right)dz \tag{10.1}$$

where,

$$S=\int_0^h\frac{z^2}{\eta}dz \tag{10.2}$$

and

$$\rho c_p\left(\frac{\partial T}{\partial t}+v_x\frac{\partial T}{\partial x}+v_y\frac{\partial T}{\partial y}\right)=\beta T\frac{\partial p}{\partial t}+\eta\dot{\gamma}^2+k\frac{\partial^2 T}{\partial z^2} \tag{10.3}$$

where

$$\dot{\gamma}^2=\sqrt{\left(\frac{\partial v_x}{\partial z}\right)^2+\left(\frac{\partial v_y}{\partial z}\right)^2} \tag{10.4}$$

Equations (10.1) and (10.3) are to be solved subject to the following boundary conditions:

i. The pressure is defined at the point(s) where the melt is injected. That is,

$$p=p_{in} \text{ at the injection point(s)} \tag{10.5}$$

ii. The normal pressure gradient is zero at any impermeable boundary. That is, at a boundary we have

$$\frac{\partial p}{\partial n}=0 \tag{10.6}$$

iii. The temperature at the cavity wall, or at some point interior to the mold wall, is specified. That is,

$$T=T_w \text{ at } z=h \text{ or } z=h+\delta \tag{10.7}$$

iv. The temperature gradient in the z-direction is zero at the cavity centre line. That is,

$$\frac{\partial T}{\partial z} = 0 \text{ at } z = 0 \tag{10.8}$$

v. The melt temperature at the point(s) of injection is defined. That is,

$$T = T_{in} \text{ at the injection point(s)} \tag{10.9}$$

Equation (10.1) differs from the equation used in the filling phase due to the additional terms on the RHS. These RHS terms account for the effect of compressibility and thermal expansion respectively. Equation (10.3) is also slightly different to that in filling as it includes the last term to account for heat due to compression.

Equation (10.1) can be written more concisely as:

$$\overline{\kappa} H \frac{\partial p}{\partial t} - (\nabla \cdot S \nabla \tilde{p}) = b \tag{10.10}$$

where ∇ denotes the gradient operator in two dimensions and

$$b = \int_0^H \frac{\beta}{\rho c_p} \left(\eta \dot{\gamma}^2 + k \frac{\partial^2 T}{\partial z^2} \right) dz \tag{10.11}$$

The thermal expansion term, b, can be calculated explicitly for given temperature and pressure fields. It can therefore be evaluated at a previous time step. Our approach then is to solve Eq. (10.10) for pressure at the new time, using the value of b at a previous time step.

The time rate of pressure change $\partial p / \partial t$, can have a large magnitude and small time steps are necessary for solution. This is particularly so at the instant the cavity is volumetrically filled or when the packing pressure is suddenly changed to simulate a profile. The rate of temperature change during packing is far lower than for pressure however. In practice performing both temperature and pressure solutions at each time step can be CPU intensive. Moldflow software has an isothermal period in which the temperature is assumed constant. The isothermal period may be defined by the user but has a default of 0.05 seconds. This means that a number of pressure solutions can be done at very small time steps during the isothermal period. This technique reduces the CPU time required for solution.

The essential steps in the packing phase solution are:

1. Calculate the pressure field at the new time t_{n+1}.
2. Calculate the temperature change due to compression of the material.
3. Calculate density and its derivatives with respect to pressure and temperature using the material's PVT data and the results of steps 1 and 2 above.
4. Update the viscosity and hence the fluidity S_2.
5. Increment the time.
6. If within the isothermal period, return to one above. If the isothermal period has expired, continue with the steps below.
7. Calculate temperature change due to compression, convection, dissipation and conduction.
8. Update viscosity.
9. Calculate pressure change due to temperature change.
10. Calculate density using PVT data.
11. Increment the time step.
12. If packing time is complete stop. If not, return to step 1 above.

10.2 Finite Element Formulation for the Pressure Field in the Cavity

The cavity is modeled with triangular elements. As in the filling phase, linear interpolation is used for the pressure within an element. Hence the pressure within element number e denoted by, $\tilde{p}^{(e)}$ is given by:

$$\tilde{p}^{(e)}(x, y) = N_1 p_1 + N_2 p_2 + N_3 p_3 \tag{10.12}$$

where the N_i are the element shape functions which, when written in terms of area coordinates (see Section 9.2.1), are:

$$
\begin{aligned}
N_1 &= \xi_1 \\
N_2 &= \xi_2 \\
N_3 &= \xi_3 \\
\xi_1 + \xi_2 + \xi_3 &= 1
\end{aligned}
\tag{10.13}
$$

Hence (10.12) becomes

$$\tilde{p}^{(e)}(\xi_1, \xi_2, \xi_3) = \xi_1 p_1 + \xi_2 p_2 + \xi_3 p_3 \tag{10.14}$$

Applying the Galerkin method to Eq. (10.10) we get the following three equations corresponding to $i = 1, 2, 3$:

$$
\begin{aligned}
&\int_A \overline{\kappa} H \frac{\partial \tilde{p}}{\partial t} N_1 dA - \int_A (\nabla \cdot S \nabla \tilde{p}) N_1 dA - \int_A b N_1 dA = 0 \\
&\int_A \overline{\kappa} H \frac{\partial \tilde{p}}{\partial t} N_2 dA - \int_A (\nabla \cdot S \nabla \tilde{p}) N_2 dA - \int_A b N_2 dA = 0 \\
&\int_A \overline{\kappa} H \frac{\partial \tilde{p}}{\partial t} N_3 dA - \int_A (\nabla \cdot S \nabla \tilde{p}) N_3 dA - \int_A b N_3 dA = 0
\end{aligned}
\tag{10.15}
$$

We first direct our attention to the first term of each of the equations in (10.15),

$$\int_A \overline{\kappa} H \frac{\partial \tilde{p}}{\partial t} N_i dA \quad i = 1, 2, 3 \tag{10.16}$$

Assuming no variation of κ or H over the area of the element, we have

$$\int_A \overline{\kappa} H \frac{\partial \tilde{p}}{\partial t} N_i dA \quad = \overline{\kappa} H \int_A \frac{\partial \tilde{p}}{\partial t} N_i dA \tag{10.17}$$

Substituting N_i and $\tilde{p}$ from equations. (10.13) and (10.14) in Eq. (10.17) we obtain:

$$\begin{aligned}\int_A \bar{\kappa} H \frac{\partial \tilde{p}}{\partial t} N_i dA &= \bar{\kappa} H \int_A \frac{\partial}{\partial t}(\xi_1 p_1 + \xi_2 p_2 + \xi_3 p_3)\xi_i dA \\ &= \bar{\kappa} H \int_A \left(\xi_1 \frac{\partial p_1}{\partial t}\xi_i + \xi_2 \frac{\partial p_2}{\partial t}\xi_i + \xi_3 \frac{\partial p_3}{\partial t}\xi_i\right) dA \\ &= \bar{\kappa} H \left\{\frac{\partial p_1}{\partial t}\int_A \xi_1\xi_i dA + \frac{\partial p_2}{\partial t}\int_A \xi_2\xi_i dA + \frac{\partial p_3}{\partial t}\int_A \xi_3\xi_i dA\right\}\end{aligned} \tag{10.18}$$

For $i = 1, 2$ and 3 we get the following equations:

$$\begin{aligned}\int_A \bar{\kappa} H \frac{\partial \tilde{p}}{\partial t} N_1 dA &= \bar{\kappa} H \left\{\frac{\partial p_1}{\partial t}\int_A \xi_1\xi_1 dA + \frac{\partial p_2}{\partial t}\int_A \xi_2\xi_1 dA + \frac{\partial p_3}{\partial t}\int_A \xi_3\xi_1 dA\right\} \\ \int_A \bar{\kappa} H \frac{\partial \tilde{p}}{\partial t} N_2 dA &= \bar{\kappa} H \left\{\frac{\partial p_1}{\partial t}\int_A \xi_1\xi_2 dA + \frac{\partial p_2}{\partial t}\int_A \xi_2\xi_2 dA + \frac{\partial p_3}{\partial t}\int_A \xi_3\xi_2 dA\right\} \\ \int_A \bar{\kappa} H \frac{\partial \tilde{p}}{\partial t} N_3 dA &= \bar{\kappa} H \left\{\frac{\partial p_1}{\partial t}\int_A \xi_1\xi_3 dA + \frac{\partial p_2}{\partial t}\int_A \xi_2\xi_3 dA + \frac{\partial p_3}{\partial t}\int_A \xi_3\xi_3 dA\right\}\end{aligned} \tag{10.19}$$

The following formula [CR] may be used to evaluate the integrals in Eq. (10.19):

$$\int_A \xi_1^k \xi_2^l \xi_3^m dA = 2A \frac{k!l!m!}{(2+k+l+m)!} \tag{10.20}$$

Using this formula, equations (10.19) become:

$$\begin{aligned}
\int_A \overline{\kappa} H \frac{\partial \tilde{p}}{\partial t} N_1 dA &= \frac{\overline{\kappa} A H}{12}\left\{2\frac{\partial p_1}{\partial t} + \frac{\partial p_2}{\partial t} + \frac{\partial p_3}{\partial t}\right\} \\
\int_A \overline{\kappa} H \frac{\partial \tilde{p}}{\partial t} N_2 dA &= \frac{\overline{\kappa} A H}{12}\left\{\frac{\partial p_1}{\partial t} + 2\frac{\partial p_2}{\partial t} + \frac{\partial p_3}{\partial t}\right\} \\
\int_A \overline{\kappa} H \frac{\partial \tilde{p}}{\partial t} N_3 dA &= \frac{\overline{\kappa} A H}{12}\left\{\frac{\partial p_1}{\partial t} + \frac{\partial p_2}{\partial t} + 2\frac{\partial p_3}{\partial t}\right\}
\end{aligned} \tag{10.21}$$

These equations can be written in matrix form as follows:

$$\left\{\begin{matrix} \overline{\kappa} H \int_A (\partial \tilde{p}/\partial t) N_1 dA \\ \overline{\kappa} H \int_A (\partial \tilde{p}/\partial t) N_2 dA \\ \overline{\kappa} H \int_A (\partial \tilde{p}/\partial t) N_3 dA \end{matrix}\right\} = \frac{\overline{\kappa} A H}{12}\begin{bmatrix} 2 & 1 & 1 \\ 1 & 2 & 1 \\ 1 & 1 & 2 \end{bmatrix}\left\{\begin{matrix} \dot{p}_1 \\ \dot{p}_2 \\ \dot{p}_3 \end{matrix}\right\} \tag{10.22}$$

where $\dot{p}_i = \partial p_i/\partial t$.

This equation may be abbreviated as follows:

$$\left\{\begin{matrix} \overline{\kappa} H \int_A (\partial \tilde{p}/\partial t) N_1 dA \\ \overline{\kappa} H \int_A (\partial \tilde{p}/\partial t) N_2 dA \\ \overline{\kappa} H \int_A (\partial \tilde{p}/\partial t) N_3 dA \end{matrix}\right\} = \left[C^{(e)}\right]\{\dot{p}\} \tag{10.23}$$

Consider now the second term in each of the three equations in (10.15). A similar term arose in the element formulation for the filling phase. Using Eq. (9.22) and the divergence theorem, we obtain

$$\begin{aligned}
\int_A (\nabla \cdot S_2 \nabla \tilde{p}) N_i &= \int_A (\nabla \cdot N_i S_2 \nabla \tilde{p}) dA - \int_A S_2(\nabla \tilde{p} \cdot \nabla N_i) dA \\
&= \int_\Sigma N_i S_2 \nabla \tilde{p} \cdot \vec{n} d\Sigma - \int_A S_2(\nabla \tilde{p} \cdot \nabla N_i) dA
\end{aligned} \tag{10.24}$$

Both terms on the RHS were evaluated in Chapter 9. where we obtained (see equations (9.32) to (9.42)) three equations, corresponding to $i = 1, 2, 3$ which could be written in the following matrix form:

$$\begin{Bmatrix} \int_A (\nabla \cdot S_2 \nabla \tilde{p}) N_i dA \\ \int_A (\nabla \cdot S_2 \nabla \tilde{p}) N_i dA \\ \int_A (\nabla \cdot S_2 \nabla \tilde{p}) N_i dA \end{Bmatrix} = \begin{Bmatrix} Q_1 \\ Q_2 \\ Q_3 \end{Bmatrix} - \begin{bmatrix} k_{11} & k_{11} & k_{11} \\ k_{21} & k_{22} & k_{23} \\ k_{31} & k_{32} & k_{33} \end{bmatrix} \begin{Bmatrix} p_1 \\ p_2 \\ p_3 \end{Bmatrix} \tag{10.25}$$

For brevity, we abbreviate this to:

$$\begin{Bmatrix} \int_A (\nabla \cdot S_2 \nabla \tilde{p}) N_i dA \\ \int_A (\nabla \cdot S_2 \nabla \tilde{p}) N_i dA \\ \int_A (\nabla \cdot S_2 \nabla \tilde{p}) N_i dA \end{Bmatrix} = \{Q\} - \left[K^{(e)}\right]\{p\} \tag{10.26}$$

Substituting equations (10.26) and (10.23) in (10.15) we obtain:

$$\left[C^{(e)}\right]\{\dot{p}\} + \left[K^{(e)}\right]\{p\} - \{Q\} + \{B\} = 0 \tag{10.27}$$

where

$$\{B\} = \begin{Bmatrix} \int_A bN_1 dA \\ \int_A bN_2 dA \\ \int_A bN_3 dA \end{Bmatrix} \tag{10.28}$$

Rearranging we get

$$\left[C^{(e)}\right]\{\dot{p}\} + \left[K^{(e)}\right]\{p\} = \{Q\} - \{B\} \tag{10.29}$$

This gives the element equations for the packing phase.

10.2.1 Assembly of Element Equations and Solution

Equation (10.29) gives the element equations to be used for the packing phase. These must be assembled in the same way as described for the filling phase (see Section 9.2.3). Denoting the system equations by similar symbols to those used above, but without the superscript we have

$$[C]\{\dot{p}\} + [K]\{p\} = \{Q\} - \{B\} \tag{10.30}$$

Using a first order forward difference approximation to the pressure derivative, and evaluating the pressure at the current time step, Eq. (10.30) may be written:

$$[C]\left(\frac{\{p\}^{n+1}-\{p\}^{n}}{\Delta t}\right)+[K]\{p\}^{n+1}=\{Q\}-\{B\} \tag{10.31}$$

After rearrangement we obtain,

$$([C]+\Delta t[K])\{p\}^{n+1}=[C]\{p\}^{n}+\Delta t\left(\{Q\}-\{B\}\right) \tag{10.32}$$

The relevant boundary conditions can then be applied and the value of $\{p\}^{n+1}$ determined.

The boundary conditions relating to $\{p\}^{n+1}$ and $\{Q\}$ are applied as in the filling phase.

10.2.2 Finite Element Pressure Solution for Runners

The pressure equation for runners in the packing phase was established in Section 6.6 and is

$$\begin{aligned}0=\overline{\kappa}_c R^2\frac{\partial p}{\partial t}-2\int_0^R\frac{\beta}{\rho c_p}\left\{\eta\dot{\gamma}^2+\frac{k}{r}\frac{\partial}{\partial r}\left(r\frac{\partial T}{\partial r}\right)\right\}r\,dr\\-\frac{\partial}{\partial x}\left(r^+ S_1\frac{\partial p}{\partial x}\right)\end{aligned} \tag{10.33}$$

Setting

$$b=-2\int_0^R\frac{\beta}{\rho c_p}\left\{\eta\dot{\gamma}^2+\frac{k}{r}\frac{\partial}{\partial r}\left(r\frac{\partial T}{\partial r}\right)\right\}r\,dr \tag{10.34}$$

this equation can be abbreviated to

$$0=\overline{\kappa}_c R^2\frac{\partial p}{\partial t}+b-\frac{\partial}{\partial x}\left(r^+ S_1\frac{\partial p}{\partial x}\right) \tag{10.35}$$

The definition of the runner element geometry is the same as for the filling phase (see Figure 9.30). As in the filling phase, linear shape functions are used to interpolate pressure between the two nodes defining the runner element. That is, the approximate solution $\tilde{p}$ has the form:

$$\begin{aligned}\tilde{p} &= \left(\frac{x_2 - x}{x_2 - x_1}\right)p_1 + \left(\frac{x - x_1}{x_2 - x_1}\right)p_2 \\ &= N_1 p_1 + N_2 p_2\end{aligned} \tag{10.36}$$

where

$$\begin{aligned}N_1 &= \frac{x_2 - x}{x_2 - x_1} = \frac{1}{L}(x_2 - x) \\ N_2 &= \frac{x - x_1}{x_2 - x_1} = \frac{1}{L}(x - x_1)\end{aligned} \tag{10.37}$$

and L is the length of the element.

A residual $R(x)$ is defined by substituting (10.36) into (10.35):

$$R(x) = \overline{\kappa}_c R^2 \frac{\partial \tilde{p}}{\partial t} + b - \frac{\partial}{\partial x}\left(r^+ S_1 \frac{\partial \tilde{p}}{\partial x}\right) \tag{10.38}$$

Galerkin's method requires that,

$$\int_{x_1}^{x_2} R(x) N_i dx = 0 \qquad i = 1, 2 \tag{10.39}$$

Substituting (10.38) into the above gives the following two equations, corresponding to $i = 1$ and $i = 2$ respectively:

$$\int_{x_1}^{x_2} R(x) N_1 dx = \overline{\kappa}_c R^2 \int_{x_1}^{x_2} \frac{\partial \tilde{p}}{\partial t} N_1 dx + \int_{x_1}^{x_2} b N_1 dx - \int_{x_1}^{x_2} \frac{\partial}{\partial x}\left(r^+ S_1 \frac{\partial \tilde{p}}{\partial x}\right) N_1 dx \tag{10.40}$$

$$\int_{x_1}^{x_2} R(x) N_2 dx = \overline{\kappa}_c R^2 \int_{x_1}^{x_2} \frac{\partial \tilde{p}}{\partial t} N_2 dx + \int_{x_1}^{x_2} b N_2 dx - \int_{x_1}^{x_2} \frac{\partial}{\partial x}\left(r^+ S_1 \frac{\partial \tilde{p}}{\partial x}\right) N_2 dx \tag{10.41}$$

Consider the first integral on the RHS of Eq. (10.40). Writing $\partial p/\partial t = \dot{p}$ we have:

$$\begin{aligned}\overline{\kappa}_c R^2 \int_{x_1}^{x_2} \frac{\partial \tilde{p}}{\partial t} N_1 dx &= \overline{\kappa}_c R^2 \int_{x_1}^{x_2} (N_1 \dot{p}_1 + N_2 \dot{p}_2) N_1 dx \\ &= \overline{\kappa}_c R^2 \left\{ \dot{p}_1 \int_{x_1}^{x_2} N_1^2 dx + \dot{p}_2 \int_{x_1}^{x_2} N_2 N_1 dx \right\}\end{aligned} \tag{10.42}$$

Using Eq. (10.37) the integrals may be evaluated:

$$\begin{aligned}\int_{x_1}^{x_2} N_1^2 dx &= \frac{1}{L^2} \int_{x_1}^{x_2} (x_2 - x)^2 dx \\ &= \frac{1}{L^2} \left[-\frac{1}{3} (x_2 - x)^3 \right]_{x=x_1}^{x=x_2} \\ &= \frac{1}{3L^2} (x_2 - x_1)^3 \\ &= \frac{L}{3}\end{aligned} \tag{10.43}$$

$$\begin{aligned}\int_{x_1}^{x_2} N_1 N_2 dx &= \frac{1}{L^2} \int_{x_1}^{x_2} (x_2 - x)(x - x_1) dx \\ &= \frac{1}{L^2} \left[\frac{1}{2} x_2 x^2 - x_2 x_1 x - \frac{1}{3} x^3 + \frac{1}{2} x^2 x_1 \right]_{x=x_1}^{x=x_2} \\ &= \frac{1}{6L^2} (x_2 - x_1)^3 \\ &= \frac{L}{6}\end{aligned} \tag{10.44}$$

Using these values in Eq. (10.42) we get:

$$\overline{\kappa}_c R^2 \int_{x_1}^{x_2} \frac{\partial \tilde{p}}{\partial t} N_1 dx = \frac{\overline{\kappa}_c L R^2}{6} (2\dot{p}_1 + \dot{p}_2) \tag{10.45}$$

Applying the same method to the first integral on the RHS of Eq. (10.41) we obtain:

$$\overline{\kappa}_c R^2 \int_{x_1}^{x_2} \frac{\partial \tilde{p}}{\partial t} N_2 dx = \frac{\overline{\kappa}_c L R^2}{6}(\dot{p}_1 + 2\dot{p}_2) \tag{10.46}$$

Consider now the last integral on the RHS of equations (10.40) and (10.41). Integrating by parts we obtain:

$$\int_{x_1}^{x_2} \frac{\partial}{\partial x}\left(r^+ S_1 \frac{\partial \tilde{p}}{\partial x}\right) N_i dx = \left(r^+ S_1 \frac{\partial \tilde{p}}{\partial x} N_i\right)\Bigg|_{x=x_1}^{x=x_2} - \int_{x_1}^{x_2} r^+ S_1 \frac{\partial \tilde{p}}{\partial x}\frac{\partial N_i}{\partial x} dx i = 1,2 \tag{10.47}$$

The terms on the RHS have already been evaluated in Section 9.2.2. Using equations (9.55) and)9.57) of that Section, we have:

$$\int_{x_1}^{x_2} \frac{\partial}{\partial x}\left(r^+ S_1 \frac{\partial \tilde{p}}{\partial x}\right) N_1 dx = q_1 - \frac{r^+ S_1}{L}(p_1 - p_2) \tag{10.48}$$

$$\int_{x_1}^{x_2} \frac{\partial}{\partial x}\left(r^+ S_1 \frac{\partial \tilde{p}}{\partial x}\right) N_2 dx = q_2 + \frac{r^+ S_1}{L}(p_1 - p_2) \tag{10.49}$$

Substituting equations (10.45) and (10.48) into (10.40) gives:

$$\int_{x_1}^{x_2} R(x) N_1 dx = \frac{\overline{\kappa} L R^2}{6}(2\dot{p}_1 + \dot{p}_2) + \int_{x_1}^{x_2} b N_1 dx - q_1 + \frac{r^+ S_1}{L}(p_1 - p_2) \tag{10.50}$$

Substituting equations (10.46) and (10.49) into (10.41) gives:

$$\int_{x_1}^{x_2} R(x) N_2 dx = \frac{\overline{\kappa} L R^2}{6}(\dot{p}_1 + 2\dot{p}_2) + \int_{x_1}^{x_2} b N_2 dx - q_2 - \frac{r^+ S_1}{L}(p_1 - p_2) \tag{10.51}$$

Equations (10.50) and (10.51) can be combined into the following matrix equation:

$$\begin{Bmatrix} \int_{x_1}^{x_2} R(x) N_1 dx \\ \int_{x_1}^{x_2} R(x) N_2 dx \end{Bmatrix} = \frac{\overline{\kappa} L R^2}{6}\begin{bmatrix} 2 & 1 \\ 1 & 2 \end{bmatrix}\begin{Bmatrix} \dot{p}_1 \\ \dot{p}_2 \end{Bmatrix} - \begin{Bmatrix} q_1 \\ q_2 \end{Bmatrix} + \frac{r^+ S_1}{L}\begin{bmatrix} 1 & -1 \\ -1 & 1 \end{bmatrix}\begin{Bmatrix} p_1 \\ p_2 \end{Bmatrix} + \begin{Bmatrix} b_1 \\ b_2 \end{Bmatrix} \tag{10.52}$$

where

$$b_1 = \int_{x_1}^{x_2} bN_1 dx \qquad b_2 = \int_{x_1}^{x_2} bN_2 dx \tag{10.53}$$

According to Eq. (10.39), the LHS of (10.52) is zero. Hence (10.52) can be rearranged as

$$\begin{aligned} \frac{\overline{\kappa} L R^2}{6} \begin{bmatrix} 2 & 1 \\ 1 & 2 \end{bmatrix} \begin{Bmatrix} \dot{p}_1 \\ \dot{p}_2 \end{Bmatrix} + \frac{r^+ S_1}{L} \begin{bmatrix} 1 & -1 \\ -1 & 1 \end{bmatrix} \begin{Bmatrix} p_1 \\ p_2 \end{Bmatrix} \\ = \begin{Bmatrix} q_1 \\ q_2 \end{Bmatrix} - \begin{Bmatrix} b_1 \\ b_2 \end{Bmatrix} \end{aligned} \tag{10.54}$$

This may be abbreviated to

$$\left[C^{(e)}\right]\{\dot{p}\} + \left[K^{(e)}\right]\{p\} = \{Q\} - \{B\} \tag{10.55}$$

This equation has identical structure to the element equations used for the cavity (see Eq. (10.29)), and is not discussed further. For details regarding the solution of this equation for p, refer to Section 10.2.1.

10.3 The Use of PVT Data and Thermal Calculations

In the packing phase, PVT data is used to determine the density of the material, the isothermal compressibility coefficient κ and the coefficient of volume expansion β.

With these quantities known, it is possible to determine the temperature increase due to compression of the material. This is given by the expression:

$$\begin{aligned} \Delta T^{n+1}_{comp} &= \Delta t \left(\frac{\beta T^n}{\rho c_p} \right) \frac{\partial p}{\partial t} \\ &= \Delta t \left(\frac{\beta T^n}{\rho c_p} \right) \frac{p^{n+1} - p^n}{\Delta t} \\ &= \frac{\beta T^n}{\rho c_p} \left(p^{n+1} - p^n \right) \end{aligned} \tag{10.56}$$

Using this temperature change, a new value of density, the isothermal compressibility coefficient κ and the coefficient of volume expansion β can be found. The derivatives

$\partial\rho/\partial p$ and $\partial\rho/\partial t$ can then be calculated from the PVT data. From these, $\partial p/\partial t$ may be found. The new values of compressibility can be used to update the matrix, $[C^{(e)}]$ of equations (10.29) and (10.56).

The increase in temperature from Eq. (10.56) is also used to update the viscosity value. This will also depend on the shear rate corresponding to the new pressure $\{p\}^{n+1}$. With the viscosity known, the fluidity S can be determined. The matrix $[K^{(e)}]$ in equations (10.29) and (10.56) is then updated.

If the time step is within an isothermal period, the time is incremented and the new pressure field is calculated using Eq. (10.56).

If it is time to start a new isothermal period (see Section 10.1), the energy equation (10.11) is solved for the temperature field. In packing the energy equation is solved in the same way as for the filling phase. The only difference is the additional source term due to compression effects. This term is the sum of the temperature changes calculated due to compression in the preceding isothermal period.

After the temperature field is calculated the viscosity, and so the fluidity, is recalculated along with new values of density, the isothermal compressibility coefficient, κ and the coefficient of volume expansion, β. The derivatives $\partial\rho/\partial p$ and $\partial\rho/\partial t$ can then be calculated from the PVT data. The new values of compressibility coefficient and fluidity can be used to update the matrices, $[C^{(e)}]$ and $[K_{(e)}]$.

The time step is now incremented and the pressure field recalculated.

Users may input both a packing time and a cooling time. During the packing time both pressure and thermal calculations are performed. At the end of packing time, the cooling time begins. During this time only thermal calculations are performed.

Appendix A
Notation Used in Text

All symbols used in the text are defined in this appendix.

A.1 Roman Symbols

c_p	Specific heat
$\hat{e}_1$, $\hat{e}_2$	Unit Vectors
$\vec{g}$	Gravitational force vector
h	Half thickness of molten region in cavity
k	Thermal conductivity
m	Mass
$\vec{n}$	Unit outward normal vector
p	Pressure
$\tilde{p}$	Approximate pressure solution
q	Volumetric flow rate
$\vec{q}$	Heat flux vector
r^+	Radius of melt channel in circular runner.
r	Radius of circular runner
t	Time

$\vec{t}$	Stress vector
$\tilde{u}(x)$	Approximate solution for which u is exact
$\vec{v}$	Velocity of fluid particle
$\overline{v}_x$	Average velocity of fluid particle in x-direction
v_x	Component of velocity in the x-direction
v_y	Component of velocity in the y-direction
v_z	Component of velocity in the z-direction
$\vec{x}$	Position vetor of fluid particle
A	Area of plane region or area of circular runner cross section
$\vec{F}$	Force vector
H	Half thickness of cavity
$\widehat{H}$	Specific enthalpy
$\underline{I}$	Unit tensor
L	Length of runner element
LHS	Left hand side
$N_i^{(e)}$	Shape function for element number
RHS	Right hand side
$R(x)$	Residual due to approximate solution
S	Boundary surface of region
$\widehat{S}$	Specific entropy

S_1	Fluidity for runner
S_2	Fluidity for cavity
T	Temperature
T_g	Glass transition temperature
$\widehat{U}$	Specific internal energy
V	Volume
$V(t)$	Material volume
$\widehat{V}$	Specific volume

A.2 Greek Symbols

α	Thermal diffusivity
β	Coefficient of volume expansion
$\dot{\gamma}$	Shear rate
$\underline{\dot{\gamma}}$	Rate of deformation tensor
$\underline{\dot{\gamma}}'$	Deviatoric rate of deformation tensor
$\dot{\gamma}_{ij}$	Component of deformation tensor
η	Viscosity function
θ	Angular displacement
κ	Isothermal compressibility coefficient

$\overline{\kappa}$	Average isothermal compressibility coefficient
λ	Time constant
μ	Newtonian viscosity
ξ	Area coordinate
ρ	Density
Σ	Boundary of plane region of area A
$\underline{\sigma}$	Stress tensor
σ_{ij}	Component of stress tensor
τ	Shear stress
$\underline{\tau}$	Viscous stress tensor
τ_{ij}	Component of viscous stress tensor

A.3 Mathematical Symbols

$\mathcal{L}$	Differential operator
∇	Gradient operator
D/Dt	Material derivative
$\int_\Sigma$	Integral over a linear boundary Σ
$\int_S$	Integral over a surface S
$\int_V$	Integral over a volume V

: Double dot tensor product

· Vector scalar product or vector tensor product

A.4 Vectors and Tensor Quantities

Vectors will be identified by an arrow above the variable. For example, $\vec{v}$.

Second order tensors are underlined. For example, $\underline{\sigma}$.

A.5 Results of Vector and Tensor Products

For multiplication of vectors and tensors, we adopt the notation of [BS]. For the dot product and double dot product of vector and tensor quantities we use the following convention:

- Operations that result in a scalar quantity will be enclosed in parentheses. For example, $(\vec{v} \cdot \vec{v})$, $(\underline{\tau} : \underline{\tau})$.
- Operations that result in a vector quantity will be enclosed in square brackets. For example, $[\nabla \cdot \underline{\tau}]$.
- Operations that result in a tensor quantity will be enclosed in curly brackets. For example, $\{\vec{v}\vec{v}\}$.

Appendix B
Vector and Tensor Operations

In this Appendix, the vector and tensor operators used in the text are explained in Cartesian and Cylindrical coordinates.

B.1 Cartesian Coordinates

B.1.1 The Gradient Operator

A rectangular cartesian coordinate system is illustrated in Figure (B.36). This coordinate system is used to describe the mold cavity.

Figure B.36: Cartesian Coordinate System

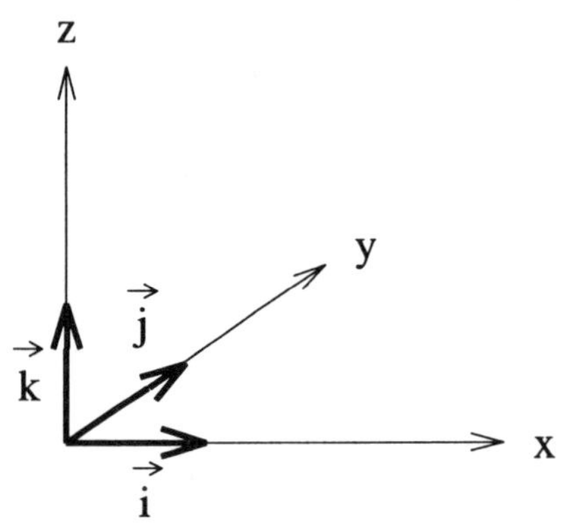

The vector differential operator ∇, is defined in cartesian coordinates as:

$$\nabla = \frac{\partial}{\partial x}\vec{i} + \frac{\partial}{\partial y}\vec{j} + \frac{\partial}{\partial z}\vec{k} \tag{B.1}$$

B.1.2 Gradient of a Scalar Function

For a scalar function $\psi = \psi(x, y, z)$ we define the gradient by:

$$\nabla\psi = \frac{\partial\psi}{\partial x}\vec{i} + \frac{\partial\psi}{\partial y}\vec{j} + \frac{\partial\psi}{\partial z}\vec{k} \tag{B.2}$$

where $\vec{i}, \vec{j}$ and $\vec{k}$ are unit vectors in the x, y and z directions as shown in Figure (B.36).

B.1.3 Laplacian of a Scalar Function

The Laplacian for a scalar field in cartesian coordinates is given by:

$$\nabla^2\psi = \frac{\partial^2\psi}{\partial x^2} + \frac{\partial^2\psi}{\partial y^2} + \frac{\partial^2\psi}{\partial z^2} \tag{B.3}$$

B.1.4 Divergence of a Vector Field

For a vector function $\vec{v} = (v_x, v_y, v_z)$ the divergence of $\vec{v}$ is defined by

$$\nabla \cdot \vec{v} = \frac{\partial v_x}{\partial x} + \frac{\partial v_y}{\partial y} + \frac{\partial v_z}{\partial z} \tag{B.4}$$

B.1.5 Gradient of a Vector Field

The gradient of a vector field $\vec{v}$ gives the tensor $\nabla\vec{v}$:

$$\begin{aligned}\nabla\vec{v} = \Bigg[&\frac{\partial v_x}{\partial x} + \frac{\partial v_y}{\partial x} + \frac{\partial v_z}{\partial x} \\ &+ \frac{\partial v_x}{\partial y} + \frac{\partial v_y}{\partial y} + \frac{\partial v_z}{\partial y} \\ &+ \frac{\partial v_x}{\partial z} + \frac{\partial v_y}{\partial z} + \frac{\partial v_z}{\partial z}\Bigg]\end{aligned} \tag{B.5}$$

The components of $\nabla\vec{v}$ are:

$$
\begin{aligned}
\{\nabla\vec{v}\}_{xx} &= \frac{\partial v_x}{\partial x} & \{\nabla\vec{v}\}_{yx} &= \frac{\partial v_x}{\partial y} & \{\nabla\vec{v}\}_{zx} &= \frac{\partial v_x}{\partial z} \\
\{\nabla\vec{v}\}_{xy} &= \frac{\partial v_x}{\partial y} & \{\nabla\vec{v}\}_{yy} &= \frac{\partial v_y}{\partial y} & \{\nabla\vec{v}\}_{zy} &= \frac{\partial v_y}{\partial z} \\
\{\nabla\vec{v}\}_{xz} &= \frac{\partial v_x}{\partial z} & \{\nabla\vec{v}\}_{yz} &= \frac{\partial v_y}{\partial z} & \{\nabla\vec{v}\}_{zz} &= \frac{\partial v_z}{\partial z}
\end{aligned}
\tag{B.6}
$$

B.1.6 Dot Product of Tensor and Vector

The dot product of a vector and tensor gives a vector. For example, $\vec{v}\cdot\nabla\vec{v}$ is a vector with the following components:

$$
\begin{aligned}
\vec{v}\cdot\nabla\vec{v} = &\left(v_x\frac{\partial v_x}{\partial x} + v_y\frac{\partial v_x}{\partial y} + v_z\frac{\partial v_x}{\partial z}\right)\vec{i} \\
&+ \left(v_x\frac{\partial v_y}{\partial x} + v_y\frac{\partial v_y}{\partial y} + v_z\frac{\partial v_y}{\partial z}\right)\vec{j} \\
&+ \left(v_x\frac{\partial v_z}{\partial x} + v_y\frac{\partial v_z}{\partial y} + v_z\frac{\partial v_z}{\partial z}\right)\vec{k}
\end{aligned}
\tag{B.7}
$$

B.1.7 Dyadic Product

The symbol $\vec{v}\vec{v}$ denotes the dyadic product of two vectors and results in a second order tensor $\underline{W}$. The components of this tensor are:

$$
\begin{aligned}
W_{xx} &= v_x v_x & W_{yx} &= v_y v_x & W_{zx} &= v_z v_x \\
W_{xy} &= v_x v_y & W_{yy} &= v_y v_y & W_{zy} &= v_z v_y \\
W_{xz} &= v_x v_z & W_{yz} &= v_y v_z & W_{zz} &= v_z v_z
\end{aligned}
\tag{B.8}
$$

B.1.8 The Rate of Deformation Tensor

The rate of deformation tensor, $\underline{\dot{\gamma}}$ is defined by

$$
\underline{\dot{\gamma}} = \nabla\vec{v} + (\nabla\vec{v})^{\dagger}
\tag{B.9}
$$

and has components

$$
\begin{aligned}
\dot{\gamma}_{xx} &= 2\frac{\partial v_x}{\partial x} & \dot{\gamma}_{xy} &= \dot{\gamma}_{yx} = \frac{\partial v_x}{\partial y} + \frac{\partial v_y}{\partial x} \\
\dot{\gamma}_{yy} &= 2\frac{\partial v_y}{\partial y} & \dot{\gamma}_{xz} &= \dot{\gamma}_{zx} = \frac{\partial v_x}{\partial z} + \frac{\partial v_z}{\partial x} \\
\dot{\gamma}_{zz} &= 2\frac{\partial v_z}{\partial z} & \dot{\gamma}_{yz} &= \dot{\gamma}_{zy} = \frac{\partial v_y}{\partial z} + \frac{\partial v_z}{\partial y}
\end{aligned}
\tag{B.10}
$$

B.1.9 Divergence of Product of a Scalar and Tensor

The divergence of the product of a scalar and a tensor gives a vector. For example,

$$
\begin{aligned}
&[\nabla \cdot (\eta \underline{\dot{\gamma}})] = [\nabla \eta \cdot \underline{\dot{\gamma}}] + \eta [\nabla \cdot \underline{\dot{\gamma}}] \\
&= \left(\frac{\partial \eta}{\partial x}\dot{\gamma}_{xx} + \frac{\partial \eta}{\partial y}\dot{\gamma}_{yx} + \frac{\partial \eta}{\partial z}\dot{\gamma}_{zx} \right)\vec{i} + \eta \left(\frac{\partial \dot{\gamma}_{xx}}{\partial x} + \frac{\partial \dot{\gamma}_{xy}}{\partial y} + \frac{\partial \dot{\gamma}_{xz}}{\partial z} \right)\vec{i} \\
&+ \left(\frac{\partial \eta}{\partial x}\dot{\gamma}_{xy} + \frac{\partial \eta}{\partial y}\dot{\gamma}_{yy} + \frac{\partial \eta}{\partial z}\dot{\gamma}_{zy} \right)\vec{j} + \eta \left(\frac{\partial \dot{\gamma}_{xy}}{\partial x} + \frac{\partial \dot{\gamma}_{yy}}{\partial y} + \frac{\partial \dot{\gamma}_{zy}}{\partial z} \right)\vec{j} \\
&+ \left(\frac{\partial \eta}{\partial x}\dot{\gamma}_{xz} + \frac{\partial \eta}{\partial y}\dot{\gamma}_{yz} + \frac{\partial \eta}{\partial z}\dot{\gamma}_{zz} \right)\vec{k} + \eta \left(\frac{\partial \dot{\gamma}_{xz}}{\partial x} + \frac{\partial \dot{\gamma}_{yz}}{\partial y} + \frac{\partial \dot{\gamma}_{zz}}{\partial z} \right)\vec{k} \\
&= \left(\frac{\partial}{\partial x}(\eta\dot{\gamma}_{xx}) + \frac{\partial}{\partial y}(\eta\dot{\gamma}_{xy}) + \frac{\partial}{\partial z}(\eta\dot{\gamma}_{xz}) \right)\vec{i} \\
&+ \left(\frac{\partial}{\partial x}(\eta\dot{\gamma}_{yx}) + \frac{\partial}{\partial y}(\eta\dot{\gamma}_{yy}) + \frac{\partial}{\partial z}(\eta\dot{\gamma}_{yz}) \right)\vec{j} \\
&+ \left(\frac{\partial}{\partial x}(\eta\dot{\gamma}_{xz}) + \frac{\partial}{\partial y}(\eta\dot{\gamma}_{yz}) + \frac{\partial}{\partial z}(\eta\dot{\gamma}_{zz}) \right)\vec{k} \\
&\quad (\text{since } \dot{\gamma}_{ij} = \dot{\gamma}_{ji}) \\
&= \left(\frac{\partial}{\partial x}\left(2\eta\frac{\partial v_x}{\partial x}\right) + \frac{\partial}{\partial y}\left(\eta\left(\frac{\partial v_x}{\partial y} + \frac{\partial v_y}{\partial x}\right)\right) + \frac{\partial}{\partial z}\left(\eta\left(\frac{\partial v_x}{\partial z} + \frac{\partial v_z}{\partial x}\right)\right)\right)\vec{i} \\
&+ \left(\frac{\partial}{\partial x}\left(\eta\left(\frac{\partial v_x}{\partial y} + \frac{\partial v_y}{\partial x}\right)\right) + \frac{\partial}{\partial y}\left(2\eta\frac{\partial v_y}{\partial y}\right) + \frac{\partial}{\partial z}\left(\eta\left(\frac{\partial v_y}{\partial z} + \frac{\partial v_z}{\partial y}\right)\right)\right)\vec{j} \\
&+ \left(\frac{\partial}{\partial x}\left(\eta\left(\frac{\partial v_x}{\partial z} + \frac{\partial v_z}{\partial x}\right)\right) + \frac{\partial}{\partial y}\left(\eta\left(\frac{\partial v_y}{\partial z} + \frac{\partial v_z}{\partial y}\right)\right) + \frac{\partial}{\partial z}\left(2\eta\frac{\partial v_z}{\partial z}\right)\right)\vec{k}
\end{aligned}
\tag{B.11}
$$

B.1.10 The Scalar Product of Two Tensors (Double Dot Product)

Two tensors $\underline{A}, \underline{B}$ multiplied together produce the following scalar result:

$$\begin{aligned} \underline{A} : \underline{B} = {} & A_{xx}B_{xx} + A_{xy}B_{yx} + A_{xz}B_{zx} \\ & + A_{yx}B_{xy} + A_{yy}B_{yy} + A_{yz}B_{zy} \\ & + A_{zx}B_{xz} + A_{zy}B_{yz} + A_{zz}B_{zz} \end{aligned} \tag{B.12}$$

B.2 Cylindrical Coordinates

B.2.1 The Gradient Operator

Figure B.37: The Cylindrical Coordinate System

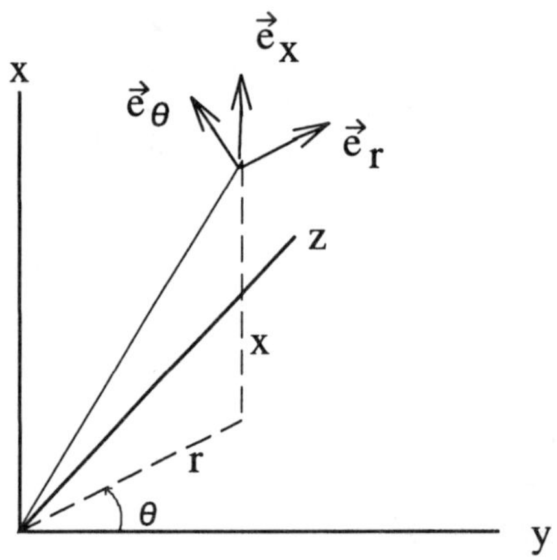

Figure B.37 illustrates a cylindrical coordinate system. This system is useful for describing fluid motion in circular runners.

The vector differential operator ∇, is defined in cylindrical coordinates as:

$$\nabla = \frac{\partial}{\partial r}\vec{e}_r + \frac{1}{r}\frac{\partial}{\partial \theta}\vec{e}_\theta + \frac{\partial}{\partial x}\vec{e}_x \tag{B.13}$$

B.2.2 Gradient of a Scalar Function

For a scalar function $\psi = \psi(r, \theta, x)$ the gradient is defined by:

$$\nabla \psi = \frac{\partial \psi}{\partial r}\vec{e}_r + \frac{1}{r}\frac{\partial \psi}{\partial \theta}\vec{e}_\theta + \frac{\partial \psi}{\partial x}\vec{e}_x \tag{B.14}$$

B.2.3 Laplacian of a Scalar Function

The Laplacian for a scalar field in cylindrical coordinates is given by:

$$\nabla^2 \psi = \nabla \cdot \nabla \psi = \frac{1}{r}\frac{\partial}{\partial r}\left(r\frac{\partial \psi}{\partial r}\right) + \frac{1}{r^2}\frac{\partial^2 \psi}{\partial \theta^2} + \frac{\partial^2 \psi}{\partial x^2} \tag{B.15}$$

B.2.4 Divergence of a Vector Field

For a vector function $\vec{v} = (v_r, v_\theta, v_x)$ the divergence of $\vec{v}$ is defined by

$$(\nabla \cdot \vec{v}) = \frac{1}{r}\frac{\partial}{\partial r}(r v_r) + \frac{1}{r}\frac{\partial v_\theta}{\partial \theta} + \frac{\partial v_x}{\partial x} \tag{B.16}$$

B.2.5 Gradient of a Vector Field

The gradient of a vector field $\vec{v}$ gives the tensor $\nabla\vec{v}$:

$$\begin{aligned}\{\nabla\vec{v}\} =& \frac{\partial v_r}{\partial r} + \frac{\partial v_\theta}{\partial r} + \frac{\partial v_x}{\partial r} \\ &+ \left(\frac{1}{r}\frac{\partial v_r}{\partial \theta} - \frac{v_\theta}{r}\right) + \left(\frac{1}{r}\frac{\partial v_\theta}{\partial \theta} + \frac{v_r}{r}\right) + \left(\frac{1}{r}\frac{\partial v_x}{\partial \theta}\right) \\ &+ \frac{\partial v_r}{\partial x} + \frac{\partial v_\theta}{\partial x} + \frac{\partial v_x}{\partial x}\end{aligned} \tag{B.17}$$

The components of the tensor are:

$$\begin{aligned}\{\nabla\vec{v}\}_{rr} &= \frac{\partial v_r}{\partial r} & \{\nabla\vec{v}\}_{\theta r} &= \frac{1}{r}\frac{\partial v_r}{\partial \theta} - \frac{v_\theta}{r} & \{\nabla\vec{v}\}_{xr} &= \frac{\partial v_r}{\partial x} \\ \{\nabla\vec{v}\}_{r\theta} &= \frac{\partial v_\theta}{\partial r} & \{\nabla\vec{v}\}_{\theta\theta} &= \frac{1}{r}\frac{\partial v_\theta}{\partial \theta} + \frac{v_r}{r} & \{\nabla\vec{v}\}_{x\theta} &= \frac{\partial v_\theta}{\partial x} \\ \{\nabla\vec{v}\}_{rx} &= \frac{\partial v_x}{\partial r} & \{\nabla\vec{v}\}_{\theta x} &= \frac{1}{r}\frac{\partial v_x}{\partial \theta} & \{\nabla\vec{v}\}_{xx} &= \frac{\partial v_x}{\partial x}\end{aligned} \tag{B.18}$$

B.2.6 Dot Product of Vector and Tensor

The dot product of a vector with a tensor is a vector. For example,

$$
\begin{aligned}
[\vec{v}\cdot\nabla\vec{v}] = &\left(v_r\frac{\partial v_r}{\partial r} + v_\theta\left(\frac{1}{r}\frac{\partial v_r}{\partial\theta} - \frac{v_\theta}{r}\right) + v_x\frac{\partial v_r}{\partial x}\right)\vec{e}_r \\
&+ \left(v_r\frac{\partial v_\theta}{\partial r} + v_\theta\left(\frac{1}{r}\frac{\partial v_\theta}{\partial\theta} + \frac{v_r}{r}\right) + v_x\frac{\partial v_\theta}{\partial x}\right)\vec{e}_\theta \\
&+ \left(v_r\frac{\partial v_x}{\partial r} + v_\theta\left(\frac{1}{r}\frac{\partial v_x}{\partial\theta}\right) + v_x\frac{\partial v_x}{\partial x}\right)\vec{e}_x
\end{aligned}
\tag{B.19}
$$

The symbol $\vec{v}\vec{v}$ is used to describe a second order tensor where the components of the tensor are formed by ordered pairs of the components of two vectors. That is $\vec{v}\vec{v}$ defines a tensor $\underline{W}$ with the following components:

$$
\begin{aligned}
W_{rr} &= v_r v_r & W_{\theta r} &= v_\theta v_r & W_{xr} &= v_x v_r \\
W_{r\theta} &= v_r v_\theta & W_{\theta\theta} &= v_\theta v_\theta & W_{x\theta} &= v_x v_\theta \\
W_{rx} &= v_r v_x & W_{\theta x} &= v_\theta v_x & W_{xx} &= v_x v_x
\end{aligned}
\tag{B.20}
$$

B.2.8 The Rate of Deformation Tensor

The rate of deformation tensor $\underline{\dot{\gamma}}$ has the following components in cylindrical coordinates:

$$
\begin{aligned}
\dot{\gamma}_{rr} &= 2\frac{\partial v_r}{\partial r} & \dot{\gamma}_{r\theta} = \dot{\gamma}_{\theta r} &= r\frac{\partial}{\partial r}\left(\frac{v_\theta}{r}\right) + \frac{1}{r}\frac{\partial v_r}{\partial\theta} \\
\dot{\gamma}_{\theta\theta} &= 2\left(\frac{1}{r}\frac{\partial v_\theta}{\partial\theta} + \frac{v_r}{r}\right) & \dot{\gamma}_{\theta x} = \dot{\gamma}_{x\theta} &= \frac{1}{r}\frac{\partial v_x}{\partial\theta} + \frac{\partial v_\theta}{\partial x} \\
\dot{\gamma}_{xx} &= 2\frac{\partial v_x}{\partial x} & \dot{\gamma}_{xr} = \dot{\gamma}_{rx} &= \frac{\partial v_x}{\partial r} + \frac{\partial v_r}{\partial x}
\end{aligned}
\tag{B.21}
$$

B.2.9 Divergence of a Tensor

The divergence of the rate of deformation tensor produces a vector with components:

$$
\begin{aligned}
\nabla\cdot(\eta\underline{\dot{\gamma}}) =& \left[\frac{1}{r}\frac{\partial}{\partial r}(r\eta\dot{\gamma}_{rr})+\frac{1}{r}\frac{\partial}{\partial\theta}(\eta\dot{\gamma}_{\theta r})+\frac{\partial}{\partial x}(\eta\dot{\gamma}_{xr})-\frac{\dot{\gamma}_{\theta\theta}}{r}\right]\vec{e}_r \\
&+\left[\frac{1}{r^2}\frac{\partial}{\partial r}(r^2\eta\dot{\gamma}_{r\theta})+\frac{1}{r}\frac{\partial}{\partial\theta}(\eta\dot{\gamma}_{\theta\theta})+\frac{\partial}{\partial x}(\eta\dot{\gamma}_{x\theta})+\eta\left(\frac{\dot{\gamma}_{\theta r}-\dot{\gamma}_{r\theta}}{r}\right)\right]\vec{e}_\theta \\
&+\left[\frac{1}{r}\frac{\partial}{\partial r}(r\eta\dot{\gamma}_{rx})+\frac{1}{r}\frac{\partial}{\partial\theta}(\eta\dot{\gamma}_{\theta x})+\frac{\partial}{\partial x}(\eta\dot{\gamma}_{xx})\right]\vec{e}_x \\
=&\left[\frac{1}{r}\frac{\partial}{\partial r}\left(2r\eta\frac{\partial v_r}{\partial r}\right)+\frac{1}{r}\frac{\partial}{\partial\theta}\left(\eta\left(r\frac{\partial}{\partial r}\left(\frac{v_\theta}{r}\right)+\frac{1}{r}\frac{\partial v_r}{\partial\theta}\right)\right)\right. \\
&\left.+\frac{\partial}{\partial x}\left(\eta\left(\frac{\partial v_x}{\partial r}+\frac{\partial v_r}{\partial x}\right)\right)-\frac{2}{r}\left(\frac{1}{r}\frac{\partial v_\theta}{\partial\theta}+\frac{v_r}{r}\right)\right]\vec{e}_r \\
&+\left[\frac{1}{r^2}\frac{\partial}{\partial r}\left(r^2\eta\left(r\frac{\partial}{\partial r}\left(\frac{v_\theta}{r}\right)+\frac{1}{r}\frac{\partial v_r}{\partial\theta}\right)\right)+\frac{1}{r}\frac{\partial}{\partial\theta}\left(2\eta\left(\frac{1}{r}\frac{\partial v_\theta}{\partial\theta}+\frac{v_r}{r}\right)\right)\right. \\
&\left.+\frac{\partial}{\partial x}\left(\eta\left(\frac{1}{r}\frac{\partial v_x}{\partial\theta}+\frac{\partial v_\theta}{\partial x}\right)\right)\right]\vec{e}_\theta \\
&+\left[\frac{1}{r}\frac{\partial}{\partial r}\left(r\eta\left(\frac{\partial v_x}{\partial r}+\frac{\partial v_r}{\partial x}\right)\right)+\frac{1}{r}\frac{\partial}{\partial\theta}\left(\eta\left(\frac{1}{r}\frac{\partial v_x}{\partial\theta}+\frac{\partial v_\theta}{\partial x}\right)\right)\right. \\
&\left.+\frac{\partial}{\partial x}\left(2\eta\frac{\partial v_x}{\partial x}\right)\right]\vec{e}_x
\end{aligned}
\tag{B.22}
$$

B.2.10 The Scalar Product of Two Tensors (Double Dot Product)

$$
\begin{aligned}
\underline{A}:\underline{B} =& A_{rr}B_{rr}+A_{r\theta}B_{\theta r}+A_{rx}B_{rx} \\
&+A_{\theta r}B_{r\theta}+A_{\theta\theta}B_{\theta\theta}+A_{\theta x}B_{x\theta} \\
&+A_{xr}B_{rx}+A_{x\theta}B_{\theta x}+A_{xx}B_{xx}
\end{aligned}
\tag{B.23}
$$

Appendix C
Dimensional Analysis of Governing Equations

In this Chapter the order of magnitude of the terms in the governing equations are estimated using characteristic values of the variables. It is shown that some terms are of such low order that they may be ignored therby simplifying the equations.

C.1 Characteristic Values

The characteristic values for the varaibles are given below:

Cavity thickness, $H = 10^{-3}\, m$
Cavity length, $L = H/\delta\, m$ where $\delta = H/L << 1$
Velocity of melt, $V = 10^{-1}\, m/s$
Cavity pressure, $p_0 = 10^7\, N/m^2$
Viscosity of melt, $\eta_0 = 10^4\, Ns/m^2$
Expansivity, $\beta_0 = 10^{-3}\, 1/K$
Thermal conductivity of melt, $k = 10^{-1}\, W/mK$
Density of melt, $\rho_0 = 10^3\, kg/m^3$
Specific heat of melt, $c_{p_0} = 10^3 J/kg\, K$
Temperature difference between mold and melt, $T_0 = 10^2\, K$
Accelaration due to gravity, $g_0 = 10 m/s^2$

Using these typical values the relevant variables in the equations can be defined in terms of dimensionless variable as follows:

x-coordinate: $x = Lx^* = [H/\delta]x^*$
y-coordinate: $y = Ly^* = [H/\delta]y^*$
z-coordinate: $z = Hz^*$
time: $t = [L/V]t^* = [H/\delta V]t^*$
x component of velocity: $v_x = [L/T]v_x^* = Vv_x^*$
y component of velocity:$v_y = [L/T]v_y^* = Vv_y^*$
z component of velocity: $v_z = [H/T]v_z^* = \delta V v_z^*$
Pressure: $p = p_0 p^*$
Expansivity: $\beta = \beta_0 \beta^*$
Viscosity: $\eta = \eta_0 \eta^*$.
Density: $\rho = \rho_0 \rho^*$

Specific heat: $c_p = c^*_{p_0}$
Temperature difference: $\Delta T = T_0 \Delta T^*$.

where the starred quantities are dimensionless and of order one.

C.2 Equations for the Filling Phase

The momentum and energy equations for the filling phase are:

$$\rho \frac{\partial \vec{v}}{\partial t} = \rho \vec{g} - \nabla p + [\nabla \cdot \eta \underline{\dot{\gamma}}] - \rho [\vec{v} \cdot \nabla \vec{v}] \tag{C.1}$$

$$\rho c_p \left(\frac{\partial T}{\partial t} + \vec{v} \cdot \nabla T \right) = \beta T \left(\frac{\partial p}{\partial t} + \vec{v} \cdot \nabla p \right) + \eta \dot{\gamma}^2 + k \nabla^2 T \tag{C.2}$$

The momentum equation (C.1) is a vector equation. When expressed in cartesian coordinates it yields the following three equations:

- x-component of momentum

$$\begin{aligned} \rho \frac{\partial v_x}{\partial t} &= \rho g_x - \frac{\partial p}{\partial x} \\ &+ \frac{\partial}{\partial x}\left(2\eta \frac{\partial v_x}{\partial x}\right) + \frac{\partial}{\partial y}\left(\eta\left(\frac{\partial v_y}{\partial x} + \frac{\partial v_x}{\partial y}\right)\right) + \frac{\partial}{\partial z}\left(\eta\left(\frac{\partial v_x}{\partial z} + \frac{\partial v_z}{\partial x}\right)\right) \\ &- \rho\left(v_x \frac{\partial v_x}{\partial x} + v_y \frac{\partial v_x}{\partial y} + v_z \frac{\partial v_x}{\partial z}\right) \end{aligned} \tag{C.3}$$

- y-component of momentum

$$\begin{aligned} \rho \frac{\partial v_y}{\partial t} &= \rho g_y - \frac{\partial p}{\partial y} \\ &+ \frac{\partial}{\partial x}\left(\eta\left(\frac{\partial v_x}{\partial y} + \frac{\partial v_y}{\partial x}\right)\right) + \frac{\partial}{\partial y}\left(2\eta \frac{\partial v_y}{\partial y}\right) + \frac{\partial}{\partial z}\left(\eta\left(\frac{\partial v_z}{\partial y} + \frac{\partial v_y}{\partial z}\right)\right) \\ &- \rho\left(v_x \frac{\partial v_y}{\partial x} + v_y \frac{\partial v_y}{\partial y} + v_z \frac{\partial v_y}{\partial z}\right) \end{aligned} \tag{C.4}$$

- z-component of momentum

$$\rho\frac{\partial v_z}{\partial t} = \rho g_z - \frac{\partial p}{\partial z} + \frac{\partial}{\partial x}\left(\eta\left(\frac{\partial v_x}{\partial z} + \frac{\partial v_z}{\partial x}\right)\right) + \frac{\partial}{\partial y}\left(\eta\left(\frac{\partial v_y}{\partial z} + \frac{\partial v_z}{\partial y}\right)\right) + \frac{\partial}{\partial z}\left(2\eta\frac{\partial v_z}{\partial z}\right) - \rho\left(v_x\frac{\partial v_z}{\partial x} + v_y\frac{\partial v_z}{\partial y} + v_z\frac{\partial v_z}{\partial z}\right) \tag{C.5}$$

Momentum: x-component

Below we have expanded out Eq. (C.3) and numbered each term.

$$\underbrace{\left(\rho\frac{\partial v_x}{\partial t}\right)}_{(i)} = \underbrace{(\rho f_x)}_{(ii)} - \underbrace{\left(\frac{\partial p}{\partial x}\right)}_{(iii)} + \underbrace{\frac{\partial}{\partial x}\left(2\eta\frac{\partial v_x}{\partial x}\right)}_{(iv)} + \underbrace{\frac{\partial}{\partial y}\left(\eta\frac{\partial v_y}{\partial x}\right)}_{(v)} + \underbrace{\frac{\partial}{\partial y}\left(\eta\frac{\partial v_x}{\partial y}\right)}_{(vi)} + \underbrace{\frac{\partial}{\partial z}\left(\eta\frac{\partial v_x}{\partial z}\right)}_{(vii)} + \underbrace{\frac{\partial}{\partial z}\left(\eta\frac{\partial v_z}{\partial x}\right)}_{(viii)} - \underbrace{\rho v_x\frac{\partial v_x}{\partial x}}_{(ix)} - \underbrace{\rho v_y\frac{\partial v_x}{\partial y}}_{(x)} - \underbrace{\rho v_z\frac{\partial v_x}{\partial z}}_{(xi)} \tag{C.6}$$

We now evaluate each term in the equation.

Term (i):

$$O\left[\rho\frac{\partial v_x}{\partial t}\right] = O\left[\rho_0\frac{V}{t}\right] = O\left[\rho_0\frac{\delta V^2}{H}\right] = O\left[10^3\frac{10^{-2}\delta}{10^{-3}}\right] = O\left[10^4\delta\right] \tag{C.7}$$

Term (ii):

$$O[\rho g_x] = O[\rho_0 g_o] = O\left[10^3 10\right] = O\left[10^4\right] \tag{C.8}$$

Term (iii):

$$O\left[\frac{\partial p}{\partial x}\right] = O\left[\frac{p_0}{L}\right] = O\left[\frac{\delta p_0}{H}\right] = O\left[\frac{10^7\delta}{10^{-3}}\right] = O[10^{10}\delta] \tag{C.9}$$

Term (iv)

$$O\left[\frac{\partial}{\partial x}\left(2\eta\frac{\partial v_x}{\partial x}\right)\right] = O\left[\frac{1}{L}\left(\eta_0\frac{V}{L}\right)\right] = O\left[\eta_0\frac{\delta^2 V}{H^2}\right] = O\left[10^4\frac{10^{-1}\delta^2}{10^{-6}}\right] = O[10^9\delta^2] \tag{C.10}$$

Term (v):

$$O\left[\frac{\partial}{\partial y}\left(\eta\frac{\partial v_y}{\partial x}\right)\right] = O\left[\frac{1}{L}\left(\eta_0\frac{V}{L}\right)\right] = O\left[\eta_0\frac{\delta^2 V}{H^2}\right] = O\left[10^4\frac{10^{-1}\delta^2}{10^{-6}}\right] = O[10^9\delta^2] \tag{C.11}$$

Term (vi):

$$O\left[\frac{\partial}{\partial y}\left(\eta\frac{\partial v_x}{\partial y}\right)\right] = O\left[\frac{1}{L}\left(\eta_0\frac{V}{L}\right)\right] = O\left[\eta_0\frac{\delta^2 V}{H^2}\right] = O\left[10^4\frac{10^{-1}\delta^2}{10^{-6}}\right] = O[10^9\delta^2] \tag{C.12}$$

Term (vii):

$$O\left[\frac{\partial}{\partial z}\left(\eta\frac{\partial v_x}{\partial z}\right)\right] = O\left[\frac{1}{H}\left(\eta_0\frac{V}{H}\right)\right] = O\left[\eta_0\frac{V}{H^2}\right] = O\left[10^4\frac{10^{-1}}{10^{-6}}\right] = O[10^9] \tag{C.13}$$

Term (viii):

$$O\left[\frac{\partial}{\partial z}\left(\eta\frac{\partial v_z}{\partial x}\right)\right] = O\left[\frac{1}{H}\left(\eta_0\frac{\delta V}{L}\right)\right] = O\left[\frac{1}{H}\eta_0\frac{\delta^2 V}{H}\right] = O\left[\frac{1}{10^{-3}}10^4\frac{10^{-1}\delta^2}{10^{-3}}\right] = O[10^9\delta^2] \tag{C.14}$$

Term (ix):

$$O\left[\rho v_x\frac{\partial v_x}{\partial x}\right] = O\left[\rho_0 V\frac{V}{L}\right] = O\left[\rho_0 V\frac{\delta V}{H}\right] = O\left[10^3 10^{-1}\frac{10^{-1}\delta}{10^{-3}}\right] = O[10^4\delta] \tag{C.15}$$

Term (x):

$$O\left[\rho v_y \frac{\partial v_x}{\partial y}\right] = O\left[\rho_0 V \frac{V}{L}\right] = O\left[\rho_0 V \frac{\delta V}{H}\right]$$
$$= O\left[10^3 10^{-1} \frac{10^{-1}\delta}{10^{-3}}\right] = O\left[10^4 \delta\right] \tag{C.16}$$

Term (xi):

$$O\left[\rho v_z \frac{\partial v_x}{\partial z}\right] = O\left[\rho_0 \delta V \frac{V}{H}\right] = O\left[10^3 10^{-1} \frac{10^{-1}\delta}{10^{-3}}\right] = O\left[10^4 \delta\right] \tag{C.17}$$

To compare these results, it is convenient to write the order under each term in Eq. (C.6):

$$\underbrace{\left(\rho \frac{\partial v_x}{\partial t}\right)}_{(10^4\delta)} = \underbrace{(\rho f_x)}_{(10^4)} - \underbrace{\left(\frac{\partial p}{\partial x}\right)}_{(10^{10}\delta)} + \underbrace{\frac{\partial}{\partial x}\left(2\eta \frac{\partial v_x}{\partial x}\right)}_{(10^9\delta^2)} + \underbrace{\frac{\partial}{\partial y}\left(\eta \frac{\partial v_y}{\partial x}\right)}_{(10^9\delta^2)}$$
$$+ \underbrace{\frac{\partial}{\partial y}\left(\eta \frac{\partial v_x}{\partial y}\right)}_{(10^9\delta^2)} + \underbrace{\frac{\partial}{\partial z}\left(\eta \frac{\partial v_x}{\partial z}\right)}_{(10^9)} + \underbrace{\frac{\partial}{\partial z}\left(\eta \frac{\partial v_z}{\partial x}\right)}_{(10^9\delta^2)} \tag{C.18}$$
$$- \underbrace{\rho v_x \frac{\partial v_x}{\partial x}}_{(10^4\delta)} - \underbrace{\rho v_y \frac{\partial v_x}{\partial y}}_{(10^4\delta)} - \underbrace{\rho v_z \frac{\partial v_x}{\partial z}}_{(10^4\delta)}$$

Momentum: y-component

Below we have expanded out Eqn. (C.4). By comparing this equation with Eq. (C.6) it is clear that the orders of magnitude of the terms are identical with those of the x-component of momentum. That is,

$$\underbrace{\left(\rho \frac{\partial v_y}{\partial t}\right)}_{(10^4\delta)} = \underbrace{(\rho f_y)}_{(10^4)} - \underbrace{\left(\frac{\partial p}{\partial y}\right)}_{(10^{10}\delta)} + \underbrace{\frac{\partial}{\partial x}\left(\eta \frac{\partial v_x}{\partial y}\right)}_{(10^9\delta^2)} + \underbrace{\frac{\partial}{\partial y}\left(\eta \frac{\partial v_y}{\partial x}\right)}_{(10^9\delta^2)}$$
$$+ \underbrace{\frac{\partial}{\partial y}\left(2\eta \frac{\partial v_y}{\partial y}\right)}_{(10^9\delta^2)} + \underbrace{\frac{\partial}{\partial z}\left(\eta \frac{\partial v_z}{\partial y}\right)}_{(10^9\delta^2)} + \underbrace{\frac{\partial}{\partial z}\left(\eta \frac{\partial v_y}{\partial z}\right)}_{(10^9)} \tag{C.19}$$
$$- \underbrace{\rho v_x \frac{\partial v_y}{\partial x}}_{(10^4\delta)} - \underbrace{\rho v_y \frac{\partial v_y}{\partial y}}_{(10^4\delta)} - \underbrace{\rho v_z \frac{\partial v_y}{\partial z}}_{(10^4\delta)}$$

Momentum: z-component

Below we have expanded out Eqn.(C.5) and numbered each term.

$$\begin{aligned}\underbrace{\left(\rho\frac{\partial v_z}{\partial t}\right)}_{(i)} &= \underbrace{(\rho g_z)}_{(ii)} - \underbrace{\left(\frac{\partial p}{\partial z}\right)}_{(iii)} + \underbrace{\frac{\partial}{\partial x}\left(\eta\frac{\partial v_x}{\partial z}\right)}_{(iv)} + \underbrace{\frac{\partial}{\partial x}\left(\eta\frac{\partial v_z}{\partial x}\right)}_{(v)} \\ &+ \underbrace{\frac{\partial}{\partial y}\left(\eta\frac{\partial v_y}{\partial z}\right)}_{(vi)} + \underbrace{\frac{\partial}{\partial y}\left(\eta\frac{\partial v_z}{\partial y}\right)}_{(vii)} + \underbrace{\frac{\partial}{\partial z}\left(2\eta\frac{\partial v_z}{\partial z}\right)}_{(viii)} \\ &- \underbrace{\rho v_x\frac{\partial v_z}{\partial x}}_{(ix)} - \underbrace{\rho v_y\frac{\partial v_z}{\partial y}}_{(x)} - \underbrace{\rho v_z\frac{\partial v_z}{\partial z}}_{(xi)}\end{aligned} \tag{C.20}$$

We now evaluate each term in the equation.

Term (i):

$$\begin{aligned}O\left[\rho\frac{\partial v_z}{\partial t}\right] &= O\left[\rho_0\frac{\delta V}{t_0}\right] = O\left[\rho_0\frac{\delta^2 V^2}{H}\right] = O\left[10^3\frac{\delta^2 10^{-2}}{10^{-3}}\right] \\ &= O[10^4\delta]\end{aligned} \tag{C.21}$$

Term (ii):

$$O[\rho g_z] = O[\rho_0 g_0] = O[10^3 10] = O[10^4] \tag{C.22}$$

Term (iii):

$$O\left[\rho\frac{\partial p}{\partial z}\right] = O\left[\frac{p_0}{H}\right] = O\left[\frac{10^7}{10^{-3}}\right] = O[10^{10}] \tag{C.23}$$

Term (iv):

$$\begin{aligned}O\left[\frac{\partial}{\partial x}\left(\eta\frac{\partial v_x}{\partial z}\right)\right] &= O\left[\frac{1}{L}\left(\eta_0\frac{V}{H}\right)\right] = O\left[\frac{\delta}{H}\left(\eta_0\frac{V}{H}\right)\right] \\ &= O\left[\frac{\delta}{10^{-3}}\left(10^4\frac{10^{-1}}{10^{-3}}\right)\right] = O[10^9\delta]\end{aligned} \tag{C.24}$$

Term (v):

$$O\left[\frac{\partial}{\partial x}\left(\eta\frac{\partial v_z}{\partial z}\right)\right]=O\left[\frac{1}{L}\left(\eta_0\frac{\delta V}{L}\right)\right]=O\left[\frac{\delta}{H}\left(\eta_0\frac{\delta^2 V}{H}\right)\right]$$
$$=O\left[\frac{\delta}{10^{-3}}\left(10^4\frac{10^{-1}\delta^2}{10^{-3}}\right)\right]=O\left[10^9\delta^3\right] \quad \textbf{(C.25)}$$

Term (vi):

$$O\left[\frac{\partial}{\partial y}\left(\eta\frac{\partial v_y}{\partial z}\right)\right]=O\left[\frac{1}{L}\left(\eta_0\frac{V}{H}\right)\right]=O\left[\frac{\delta}{H}\left(\eta_0\frac{V}{H}\right)\right]$$
$$=O\left[\frac{\delta}{10^{-3}}\left(10^4\frac{10^{-1}}{10^{-3}}\right)\right]=O\left[10^9\delta\right] \quad \textbf{(C.26)}$$

Term (vii):

$$O\left[\frac{\partial}{\partial y}\left(\eta\frac{\partial v_z}{\partial y}\right)\right]=O\left[\frac{1}{L}\left(\eta_0\frac{\delta V}{L}\right)\right]=O\left[\frac{\delta}{H}\left(\eta_0\frac{\delta^2 V}{H}\right)\right]$$
$$=O\left[\frac{\delta}{10^{-3}}\left(10^4\frac{10^{-1}\delta^2}{10^{-3}}\right)\right]=O\left[10^9\delta^3\right] \quad \textbf{(C.27)}$$

Term (viii):

$$O\left[\frac{\partial}{\partial z}\left(2\eta\frac{\partial v_z}{\partial z}\right)\right]=O\left[\frac{1}{H}\left(\eta_0\frac{\delta V}{H}\right)\right]=O\left[\frac{1}{10^{-3}}\left(10^4\frac{10^{-1}\delta}{10^{-3}}\right)\right]$$
$$=O\left[10^9\delta\right] \quad \textbf{(C.28)}$$

Term (ix):

$$O\left[\rho v_x\frac{\partial v_z}{\partial x}\right]=O\left[\rho_0 V\frac{\delta V}{L}\right]=O\left[\rho_0 V\frac{\delta^2 V}{H}\right]=O\left[10^3 10^{-1}\frac{10^{-1}\delta^2}{10^{-3}}\right]$$
$$=O\left[10^4\delta^2\right] \quad \textbf{(C.29)}$$

Term (x):

$$O\left[\rho v_y\frac{\partial v_z}{\partial y}\right]=O\left[\rho_0 V\frac{\delta V}{L}\right]=O\left[\rho_0 V\frac{\delta^2 V}{H}\right]=O\left[10^3 10^{-1}\frac{10^{-1}\delta^2}{10^{-3}}\right]$$
$$=O\left[10^4\delta^2\right] \quad \textbf{(C.30)}$$

Term (xi):

$$O\left[\rho v_z \frac{\partial v_z}{\partial z}\right] = O\left[\rho_0 \delta V \frac{\delta V}{H}\right] = O\left[10^3 \delta 10^{-1} \frac{10^{-1}\delta}{10^{-3}}\right] \\ = O\left[10^4 \delta^2\right] \tag{C.31}$$

To compare these magnitudes, it is convenient to write the order under each term:

$$\underbrace{\left(\rho \frac{\partial v_z}{\partial t}\right)}_{(10^4\delta^2)} = \underbrace{(\rho g_z)}_{(10^4)} - \underbrace{\left(\frac{\partial p}{\partial z}\right)}_{(10^{10})} + \underbrace{\frac{\partial}{\partial x}\left(\eta \frac{\partial v_x}{\partial z}\right)}_{(10^9\delta)} + \underbrace{\frac{\partial}{\partial x}\left(\eta \frac{\partial v_z}{\partial x}\right)}_{(10^9\delta^3)} \\ + \underbrace{\frac{\partial}{\partial y}\left(\eta \frac{\partial v_y}{\partial z}\right)}_{(10^9\delta)} + \underbrace{\frac{\partial}{\partial y}\left(\eta \frac{\partial v_z}{\partial y}\right)}_{(10^9\delta^3)} + \underbrace{\frac{\partial}{\partial z}\left(2\eta \frac{\partial v_z}{\partial z}\right)}_{(10^9\delta)} \\ - \underbrace{\rho v_x \frac{\partial v_z}{\partial x}}_{(10^4\delta^2)} - \underbrace{\rho v_y \frac{\partial v_z}{\partial y}}_{(10^4\delta^2)} - \underbrace{\rho v_z \frac{\partial v_z}{\partial z}}_{(10^4\delta^2)} \tag{C.32}$$

The Energy Equation for the Cavity

The energy equation for the filling phase Eq. (5.12) may be written in Cartesian coordinates as follows:

$$\rho c_p\left(\frac{\partial T}{\partial t} + v_x \frac{\partial T}{\partial x} + v_y \frac{\partial T}{\partial y} + v_z \frac{\partial T}{\partial z}\right) = \beta T\left(\frac{\partial p}{\partial t} + v_x \frac{\partial p}{\partial x} + v_y \frac{\partial p}{\partial y} + v_z \frac{\partial p}{\partial z}\right) \\ + \eta\dot{\gamma}^2 + k\left(\frac{\partial^2 T}{\partial x^2} + \frac{\partial^2 T}{\partial y} + \frac{\partial^2 T}{\partial z}\right) \tag{C.33}$$

The energy equation (C.2) may be written in Cartesian coordinates as follows:

$$\rho c_p\left(\frac{\partial T}{\partial t} + v_x \frac{\partial T}{\partial x} + v_y \frac{\partial T}{\partial y} + v_z \frac{\partial T}{\partial z}\right) = \beta T\left(\frac{\partial p}{\partial t} + v_x \frac{\partial p}{\partial x} + v_y \frac{\partial p}{\partial y} + v_z \frac{\partial p}{\partial z}\right) \\ + \eta\dot{\gamma}^2 + k\left(\frac{\partial^2 T}{\partial x^2} + \frac{\partial^2 T}{\partial y} + \frac{\partial^2 T}{\partial z}\right) \tag{C.34}$$

As for the momentum equations, we estimate the order of magnitude of each term using characteristic values. Expanding the equation and numbering each of the terms we obtain:

$$\underbrace{\rho c_p \frac{\partial T}{\partial t}}_{(i)} + \underbrace{\rho c_p v_x \frac{\partial T}{\partial x}}_{(ii)} + \underbrace{\rho c_p v_y \frac{\partial T}{\partial y}}_{(iii)} + \underbrace{\rho c_p v_z \frac{\partial T}{\partial z}}_{(iv)}$$
$$= \underbrace{\beta T \frac{\partial p}{\partial t}}_{(v)} + \underbrace{\beta T v_x \frac{\partial p}{\partial x}}_{(vi)} + \underbrace{\beta T v_y \frac{\partial p}{\partial y}}_{(vii)} + \underbrace{\beta T v_z \frac{\partial p}{\partial z}}_{(viii)}$$
$$+ \underbrace{\eta \dot{\gamma}^2}_{(ix)} + \underbrace{k \frac{\partial^2 T}{\partial x^2}}_{(x)} + \underbrace{k \frac{\partial^2 T}{\partial y}}_{(xi)} + \underbrace{k \frac{\partial^2 T}{\partial z}}_{(xii)} \tag{C.35}$$

We now evaluate the order of magnitude of each of the terms.

Term (i):

$$O\left[\rho c_p \frac{\partial T}{\partial t}\right] = O\left[\rho_0 c_{p_0} \frac{T_0}{t_0}\right] = O\left[\rho_0 c_{p_0} \frac{\delta T_0 V}{H}\right] = O\left[10^3 10^3 \frac{10^2 10^{-1} \delta}{10^{-3}}\right]$$
$$= O\left[10^{10} \delta\right] \tag{C.36}$$

Term (ii):

$$O\left[\rho c_p v_x \frac{\partial T}{\partial x}\right] = O\left[\rho_0 c_{p_0} V \frac{T_0}{L}\right] = O\left[\rho_0 c_{p_0} V \frac{\delta T_0}{H}\right] = O\left[10^3 10^3 10^{-1} \frac{10^2 \delta}{10^{-3}}\right]$$
$$= O\left[10^{10} \delta\right] \tag{C.37}$$

Term (iii):

$$O\left[\rho c_p v_y \frac{\partial T}{\partial y}\right] = O\left[\rho_0 c_{p_0} V \frac{T_0}{L}\right] = O\left[\rho_0 c_{p_0} V \frac{\delta T_0}{H}\right] = O\left[10^3 10^3 10^{-1} \frac{10^2 \delta}{10^{-3}}\right]$$
$$= O\left[10^{10} \delta\right] \tag{C.38}$$

Term (iv):

$$O\left[\rho c_p v_z \frac{\partial T}{\partial z}\right] = O\left[\rho_0 c_{p_0} \delta V \frac{T_0}{H}\right] = O\left[10^3 10^3 \delta 10^{-1} \frac{10^2}{10^{-3}}\right]$$
$$= O\left[10^{10} \delta\right] \tag{C.39}$$

Term (v):

$$O\left[\beta T\frac{\partial p}{\partial t}\right]=O\left[\beta_0T_0\frac{p_0}{t_0}\right]=O\left[\beta_0T_0\frac{p_0\delta V}{H}\right]=O\left[10^{-3}10^{2}\frac{10^{7}\delta10^{-1}}{10^{-3}}\right]=O\left[10^{8}\delta\right] \tag{C.40}$$

Term (vi):

$$\begin{aligned}O\left[\beta Tv_x\frac{\partial p}{\partial x}\right]&=O\left[\beta_0T_0V\frac{p_0}{L}\right]=O\left[\beta_0T_0V\frac{\delta p_0}{H}\right]\\&=O\left[10^{-3}10^{2}10^{-1}\frac{10^{7}\delta}{10^{-3}}\right]=O\left[10^{8}\delta\right]\end{aligned} \tag{C.41}$$

Term (vii):

$$\begin{aligned}O\left[\beta Tv_y\frac{\partial p}{\partial y}\right]&=O\left[\beta_0T_0V\frac{p_0}{L}\right]=O\left[\beta_0T_0V\frac{\delta p_0}{H}\right]\\&=O\left[10^{-3}10^{2}10^{-1}\frac{10^{7}\delta}{10^{-3}}\right]=O\left[10^{8}\delta\right]\end{aligned} \tag{C.42}$$

Term (viii):

$$\begin{aligned}O\left[\beta Tv_z\frac{\partial p}{\partial z}\right]&=O\left[\beta_0T_0\delta V\frac{p_0}{H}\right]=O\left[10^{-3}10^{2}\delta10^{-1}\frac{10^{7}}{10^{-3}}\right]\\&=O\left[10^{8}\delta\right]\end{aligned} \tag{C.43}$$

Term (ix)

$$\begin{aligned}O\left[\eta\dot{\gamma}^2\right]&=O\left[\eta_0\left(\left(\frac{\partial v_x}{\partial z}\right)^2+\left(\frac{\partial v_y}{\partial z}\right)^2\right)\right]\\&=O\left[\eta_0\left(\frac{V}{H}\right)^2\right]O\left[10^{4}\left(\frac{10^{-1}}{10^{-3}}\right)^2\right]\\&=O\left[10^{8}\right]\end{aligned} \tag{C.44}$$

Term (x):

$$\begin{aligned}O\left[k\frac{\partial^2T}{\partial x^2}\right]&=O\left[k_0\frac{T_0}{L^2}\right]=O\left[k_0\frac{T_0\delta^2}{H^2}\right]=O\left[10^{-1}\frac{10^{2}\delta^2}{10^{-6}}\right]\\&=O\left[10^{7}\delta^2\right]\end{aligned} \tag{C.45}$$

Term (xi):

$$O\left[k\frac{\partial^2 T}{\partial y^2}\right] = O\left[k_0\frac{T_0}{L^2}\right] = O\left[k_0\frac{T_0\delta^2}{H^2}\right] = O\left[10^{-1}\frac{10^2\delta^2}{10^{-6}}\right] \\ = O\left[10^7\delta^2\right] \tag{C.46}$$

Term (xii):

$$O\left[k\frac{\partial^2 T}{\partial z^2}\right] = O\left[k_0\frac{T_0}{H^2}\right] = O\left[10^{-1}\frac{10^2}{10^{-6}}\right] \\ = O\left[10^7\right] \tag{C.47}$$

For convenience we write these magnitudes under their respective terms below:

$$\underbrace{\rho c_p\frac{\partial T}{\partial t}}_{(10^{10}\delta)} + \underbrace{\rho c_p v_x\frac{\partial T}{\partial x}}_{(10^{10}\delta)} + \underbrace{\rho c_p v_y\frac{\partial T}{\partial y}}_{(10^{10}\delta)} + \underbrace{\rho c_p v_z\frac{\partial T}{\partial z}}_{(10^{10}\delta)} \\ = \underbrace{\beta T\frac{\partial p}{\partial t}}_{(10^8\delta)} + \underbrace{\beta T v_x\frac{\partial p}{\partial x}}_{(10^8\delta)} + \underbrace{\beta T v_y\frac{\partial p}{\partial y}}_{(10^8\delta)} + \underbrace{\beta T v_z\frac{\partial p}{\partial z}}_{(10^8\delta)} \\ + \underbrace{\eta\dot{\gamma}^2}_{(10^8)} + \underbrace{k\frac{\partial^2 T}{\partial x^2}}_{(10^7\delta^2)} + \underbrace{k\frac{\partial^2 T}{\partial y}}_{(10^7\delta^2)} + \underbrace{k\frac{\partial^2 T}{\partial z}}_{(10^7)} \tag{C.48}$$

Appendix D
References

[AC] Austin, C.A., Filling of Mold Cavities in *Computer Aided Engineering for Injection Molding* Ed. E.C. Bernhardt, Hanser, New York, (1983).

[AR] Aris, R. *Vectors, Tensors and the Basic Equations of Fluid Mechanics*, Dover, New York, (1989).

[AW] Ames, W.F., *Numerical Methods for Partial Differential Equations 2nd Ed* Academic Press, New York, (1977).

[BA] Bird, R.B., Armstrong, R.C., and Hassager, O., *Dynamics of Polymeric Liquids: Volume 1 Fluid Mechanics,* John Wiley & Sons, New York, (1977).

[BD] Burnett, David. S. *Finite Element Analysis: From Concepts to Applications.* Addison-Wesley, New York, (1987).

[BI] Barrie, I.T., Understanding How an Injection Mold Fills, *SPE Journal*, 64-69, Aug 1971.

[BG] Broyer, E., Gutfinger, C., and Tadmor, Z., A Theoretical Model for the Cavity Filling Process in Injection Molding, *Trans. of the Society of Rheology, 19,* 423-444 (1975)

[BK] Bathe, K.J., *Finite Element Procedures in Engineering Analysis*, Prentice-Hall, Englewood Cliffs, NJ, (1982).

[BL] Bird, R.A., Stewart, W.E. and Lightfoot, E.N., *Transport Phenomena*, John Wiley & Sons, New York, (1960).

[BM] Berins, M.L. (Ed.), *Plastics Engineering Handbook of the Society of the Plastics Industry 5th Ed.,* Van Nostrand Reinhold, New York, (1991)

[BS] van den Brule, B.H.A.A. and Slikkerveer, P.J., Anisotropic Conduction of Heat Caused by Molecular Orientation in a Flowing Polymeric Liquid, *Rheol. Acta* 29, 175-181, (1990).

[BW] Boshouwers, G., and van der Werf, J., *Inject-3 a Simulation Code for the Filling Stage of the Injection Moulding Process of Thermoplastics*, Doctoral Dissertation, Technische Universiteit Eindhoven, (1988).

[CF] Friedl, C., Austin,C.A., and Thomas, R. Melt Compressibility and Expansivity in Mold Filling Simulation. *SPE Annual Technical Conference Proceedings, Vol XXXVI*(1990).

[CH] Chiang, H.H., Hieber, C.A. and Wang, K.K., A Unified Simulation of the Filling and Postfilling Stages in Injection Molding. Part II: Experimental Verification, *Polymer Eng. Sci.*, 31, 125–139, (1991).

[CR] Cook, R.D. *Concepts and Applications of Finite Element Analysis (3rd Ed)* John Wiley & Sons, New York, (1989).

[FA] Fredrickson, A.G., *Principles and Applications of Rheology*, Prentice Hall, NJ. (1964).

[FC] Fletcher, C.A.J *Computational Techniques for Fluid Dynamics Vols 1 & 2*, Springer Verlag, Berlin, Heidelberg (1988).

[GB] Gahleitner, M., Bihlmayer, G. and Sobczak, R., Modified Carreau Model, *Kunststoffe German Plastics*, 81, 54–55, (1991).

[GS] Guceri, S.I., Finite Difference Solution of Field Problems in *Computer Modeling for Polymer Processing* Ed. Tucker C.L., Hanser (1989).

[HL] Hageman, L.J., *Recent Advances in Injection Molding Simulation,* Society of Plastics Engineers, RETEC Boston, November 1990.

[HS] Hieber, C.A. and Shen, S.F. A Finite Element/Finite Difference Simulation of the Injection-Molding Filling Process, *J. Non-Newtonian Fluid Mech.*, 7, 1-32 (1980).

[HT] Huebner, K.H. and Thornton, E.A.,*The Finite Element Method for Engineers (2nd Ed)* John Wiley & Sons, New York, (1982).

[HU] Hughes, T.J.R., *The Finite Element Method: Linear Static and Dynamic Finite Element Analysis*, Prentice Hall Inc., New Jersey, (1987).

[HY] Hageman, L.A., and Young, D.M., *Applied Iterative Methods* Academic Press Inc., San Diego, (1981).

[KK] Kamal, M.R., and Kenig, S. The Injection Molding of Thermoplastics Part 1: Theoretical Model, *Polym. Eng. Sci.*, 12, 295-308 (1972).

[LC] Lee, C.C., and Castro, J.M., Model Simplification in *Fundamentals of computer modeling for Polymer Processing* Ed. C.L. Tucker III Hanser Publishers, Munich, Vienna, New York (1989).

[LW] Lord, H.A., and Williams, G., Mold Filling Studies for the Injection Molding of Thermoplastic Materials Part II: The Transient Flow of Plastic Materials in the Cavities of Injection-Molding Dies *Polymer Eng. Sci.*, 15, 569-582 (1975).

[ML] Manzione, L.T., (Ed.) *Applications of Computer Aided Engineering in Injection Molding*, Hanser, New York, (1987).

[MM] Menges, G. and Mohren, P., *How to Make Injection Molds*, Hanser, New York, (1986).

[MT] Marsden, J.E., and Tromba, A.J., *Vector Calculus*, W. H. Freeman and Company, San Francisco, (1976).

[RD] Rheinfeld, D. Influencing Moulding Quality During Injection Molding in *Injection Molding Technology*, VDI-Verlag, Dusseldorf (1981).

[RF] Rodriguez, F., *Principles of Polymer Systems*, McGraw-Hill, New York, (1982).

[RI] Rubin, I.I., *Injection Molding Theory and Practice*, John Wiley & Sons, New York, (1972).

[RR] Rosato, D.V., and Rosato, D.V., (Eds.) *Injection Molding Handbook*, Van Nostrand Reinhold, New York, (1986).

[SG] Smith, G.D., *Numerical Solution of Partial Differential Equations: Finite Difference Methods 3rd Ed.*, Oxford University Press, (1985).

[SH] Shames, I.H., *Introduction to Solid Mechanics (2nd Ed)* Prentice Hall, NJ. (1989).

[SJ] Stevenson, J.F. et al., An Experimental Study and Simulation of Disk Filling by Injection Molding, *Soc. Plastics Engrs. Tech. Papers*, 22, 282-288 (1976).

[SS] Sears, F.W., Salinger, G.L., *Thermodynamics, Kinetic Theory and Statistical Thermodynamics (3rd Ed)* Addison-Wesley Publishing Company, New York, (1975).

[TB] Tadmor, Z., Broyer, E., and Gutfinger, C., Flow Analysis Network (FAN)- A Method for Solving Flow Problems in Polymer Processing, *Polym. Eng. Sci.*, 14, 660-665 (1974).

[TC] Tucker, C.L., (Ed.) *Computer Modeling for Polymer Processing* Hanser (1989).

[WD] van Wijngaarden, H., Dijksman, J.F., Wesseling, P., Non-Isothermal Flow of a Molten Polymer in a Narrow Rectangular Cavity, *J. Non-Newtonian Fluid Mechanics,* 11, 175-199, (1982).

[WL] Williams, G., and Lord, H.A., Mold Filling Studies for the Injection Molding of Thermoplastic Materials Part I: The Flow of Plastic Materials in Hot and Cold Walled Circular Channels. *Polymer Eng. Sci.*, 15, 553-568 (1975).

[WP] Wippenbeck, P. The Mould Filling Process - Technical Requirements and Findings in *Injection Molding Technology*, VDI-Verlag, Dusseldorf (1981).

[YG} Young, D.M., and Gregory, R.T.,*A Survey of Numerical Mathematics Vol II*, Dover, New York, (1988)

Index

D

E

F

G

H

R

T